Student Solutions Manua

Introduction to Statistics & Data Analysis

FOURTH EDITION

Roxy Peck
California Polytechnic State University,
San Luis Obispo

Chris Olsen
Thomas Jefferson High School,
Cedar Rapids, Iowa

Jay Devore
California Polytechnic State University,
San Luis Obispo

Prepared by

Michael Allwood
Bruswick School, Greenwich, CT

Australia • Brazil • Japan • Korea • Mexico • Singapore • Spain • United Kingdom • United States

For product information and technology assistance, contact us at
Cengage Learning Customer & Sales Support,
1-800-354-9706

For permission to use material from this text or product, submit all requests online at **www.cengage.com/permissions**
Further permissions questions can be emailed to
permissionrequest@cengage.com

ISBN-13: 978-0-8400-6840-8
ISBN-10: 0-8400-6840-9

Brooks/Cole
20 Channel Center Street
Boston, MA 02210
USA

Cengage Learning is a leading provider of customized learning solutions with office locations around the globe, including Singapore, the United Kingdom, Australia, Mexico, Brazil, and Japan. Locate your local office at:
www.cengage.com/global

Cengage Learning products are represented in Canada by Nelson Education, Ltd.

To learn more about Brooks/Cole, visit
www.cengage.com/brookscole

Purchase any of our products at your local college store or at our preferred online store
www.cengagebrain.com

Printed in the United States of America
1 2 3 4 5 6 7 15 14 13 12 11

A Note About Rounding

In this solutions manual, rounded values are written in the calculations, but more accurate values were used to arrive at the given answers.

So, for example, a calculation in the solution to Exercise 13.59, Part (a), is shown as

$$t = \frac{-0.621}{\sqrt{\frac{1-(-0.621)^2}{59}}} = -6.090.$$

When we calculate the value of t using the numbers shown, we get an answer of −6.086, not −6.090. However, the value of −0.621 was actually, more accurately, −0.6212889827, and it was this value that was used in the calculation, giving −6.090, which is the correct value of t to the nearest one-thousandth.

Michael Allwood
Brunswick School
Greenwich, CT

Table of Contents

Chapter 1
The Role of Statistics and the Data Analysis Process

1.1 *Descriptive statistics* is the branch of statistics that involves the organization and summary of the values in a data set. *Inferential statistics* is the branch of statistics concerned with reaching conclusions about a population based on the information provided by a sample.

1.3 The percentages would have been computed from a sample.

1.5 The population of interest is the set of all 15,000 students at the university. The sample is the two hundred students who are interviewed.

1.7 The population is the set of all 7000 property owners. The sample is the 500 owners included in the survey.

1.9 The population is the set of 5000 used bricks. The sample is the set of 100 bricks she checks.

1.11 **a** The researchers wanted to compare the effectiveness of the new flu vaccine (administered by nasal spray) with the effectiveness of the conventional vaccine (administered by injection). They were motivated to learn whether the new vaccine significantly reduced the incidence of influenza (when compared to a placebo) and whether the incidence of ear infections would be reduced in those children who *did* contract influenza.

b First, it is not stated whether the subjects in the experiment were randomly assigned to the treatments; this would be necessary in a well designed experiment. Second, in order to compare the effectiveness of the new and old vaccines, it might have been useful to include a group of subjects who are given the conventional vaccine (although the results of previous studies could possibly be used for this purpose). Third, in order to determine whether the new vaccine significantly reduced the incidence of ear infections, a larger number of subjects needed to be included in the group of subjects who were given the new vaccine. With just one percent of the 1070 subjects contracting influenza (approximately 11 subjects), it is not possible to make, with a reasonable degree of confidence, an accurate estimate of the proportion of flu contractors who go on to contract the ear infection.

1.13 **a** Categorical

b Categorical

c Numerical (discrete)

d Numerical (continuous)

e Categorical

f Numerical (continuous)

1.15 **a** Continuous

b Continuous

c Continuous

d Discrete

1.17 **a** Gender of purchaser, brand of motorcycle, telephone area code

b Number of previous motorcycles

c Bar chart

d Dotplot

1.19 **a**

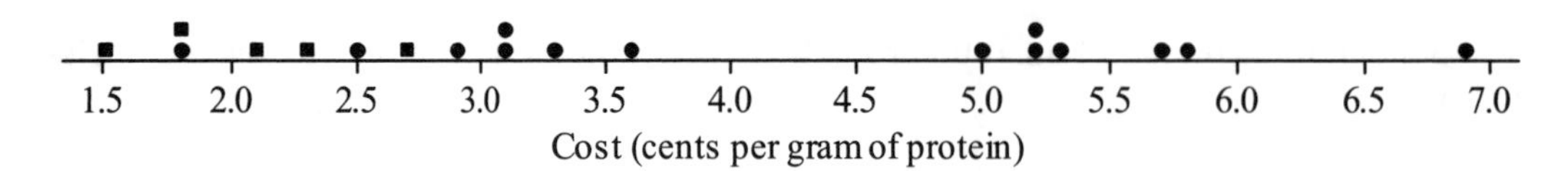

b The costs per gram of protein for the meat and poultry items are represented by squares in the dotplot above. With every one of the meat and poultry items included in the lowest seven cost per gram values, meat and poultry items appear to be relatively low cost sources of protein.

1.21

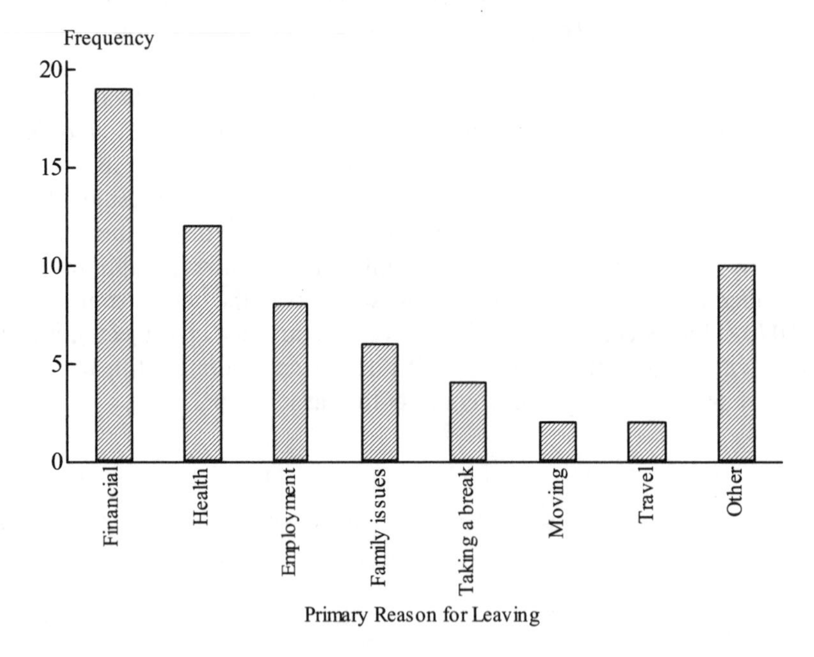

The most common reason was financial, this accounting for 30.2% of students who left for non-academic reasons. The next two most common reasons were health and other personal reasons, these accounting for 19.0% and 15.9%, respectively, of the students who left for non-academic reasons.

1.23 **a** The dotplot shows that there were two sites that received far greater numbers of visits than the remaining 23 sites. Also, it shows that the distribution of the number of visits has the greatest density of points for the smaller numbers of visits, with the density decreasing as the number of visits increases. This is the case even when only the 23 less popular sites are considered.

b Again, it is clear from the dotplot that there were two sites that were used by far greater numbers of individuals (unique visitors) than the remaining 23 sites. However, these two sites are less far above the others in terms of the number of unique visitors than they are in terms of the total number of visits. As with the distribution of the total number of visits, the distribution of the number of unique visitors has the greatest density of points for the smaller numbers of visitors, with the density decreasing as the number of unique visitors increases. This is the case even when only the 23 less popular sites are considered.

c The statistic "visits per unique visitor" tells us how heavily the individuals are using the sites. Although the table tells us that the most popular site (Facebook) in terms of the other two statistics also has the highest value of this statistic, the dotplot of visits per unique visitor shows that no one or two individual sites are far ahead of the rest in this respect.

1.25 **a**

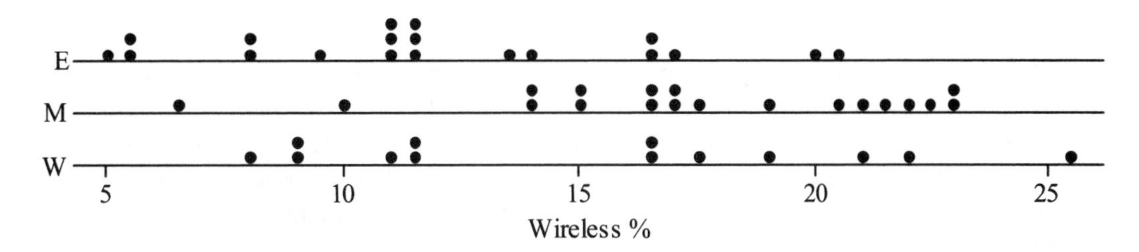

b Looking at the dotplot we can see that Eastern states have, on average, lower wireless percents than states in the other two regions. The West and Middle states regions have, on average, roughly equal wireless percents.

1.27 **a** When ranking the airlines according to delayed flights, one airline would be ranked above another if the probability of a randomly chosen flight being delayed is smaller for the first airline than it is for the second airline. These probabilities are estimated using the *rate per 10,000 flights* values, and so these are the data that should be used for this ranking. (Note that the *total number of flights* values are not suitable for this ranking. Suppose that one airline had a larger number of delayed flights than another airline. It is possible that this could be accounted for merely through the first airline having more flights than the second.)

b There are two airlines, ExpressJet and Continental, which, with 4.9 and 4.1 of every 10,000 flights delayed, stand out as the worst airlines in this regard. There are two further airlines that stand out above the rest: Delta and Comair, with rates of 2.8 and 2.7 delayed flights per 10,000 flights. All the other airlines have rates below 1.6, with the best rating being for Southwest, with a rate of only 0.1 delayed flights per 10,000.

1.29 **a**

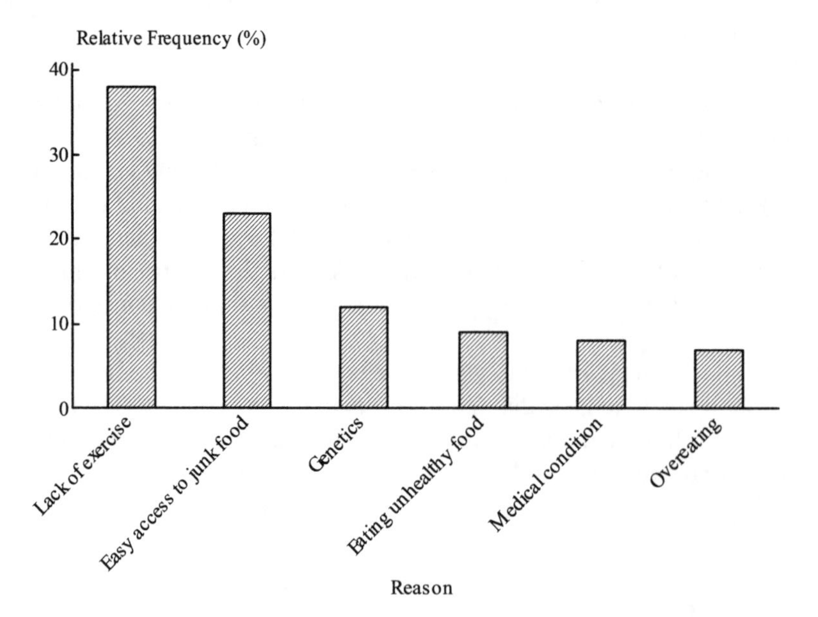

b The categories "Easy access to junk food," "Eating unhealthy food," and "Overeating" could be combined, since these categories all concern the child's eating habits. It could be considered a good idea to do this since the other three categories represent very distinct causes of the overweight condition, while for many children with poor eating habits the choice might be somewhat arbitrary as to which of the three dietary factors should be considered the most important.

1.31

Type of Household	Relative Frequency
Nonfamily	0.29
Married with children	0.27
Married without children	0.29
Single parent	0.15

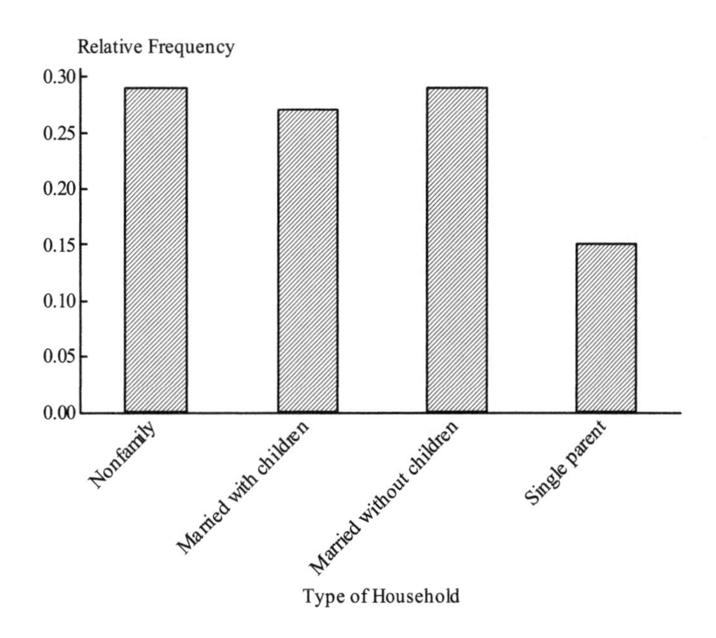

1.33 **a** Categorical

b No, since dotplots are used for numerical data.

c

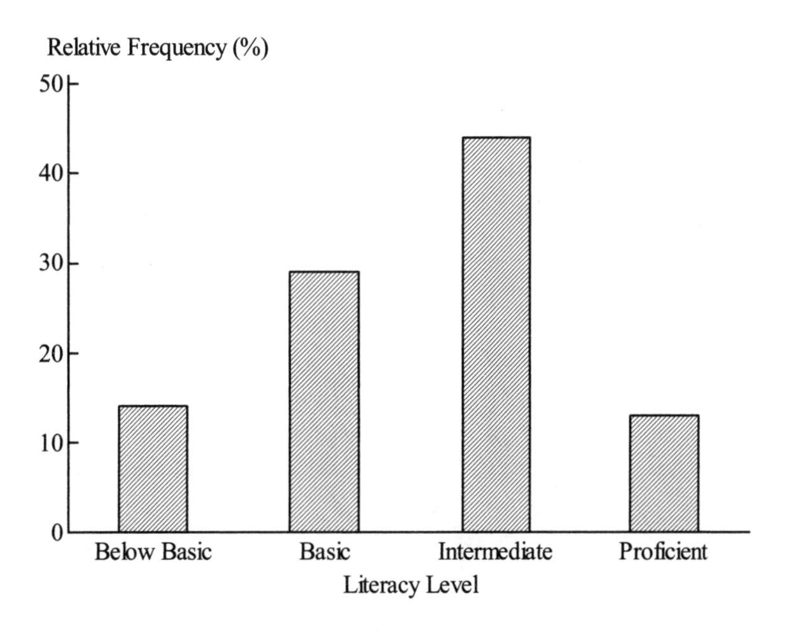

1.35

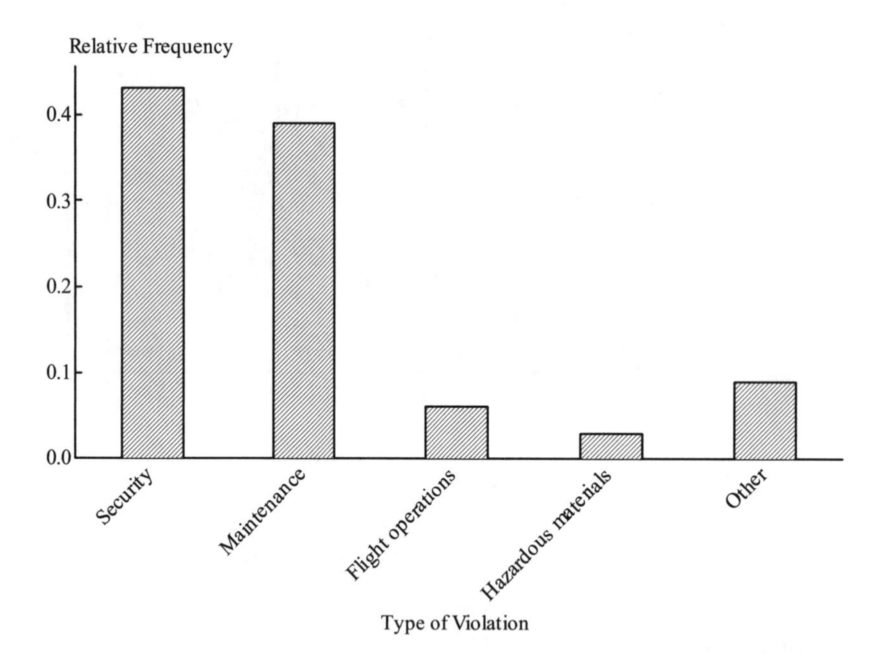

By far the most frequently occurring violation categories were security (43%) and maintenance (39%). The least frequently occurring violation categories were flight operations (6%) and hazardous materials (3%).

1.37

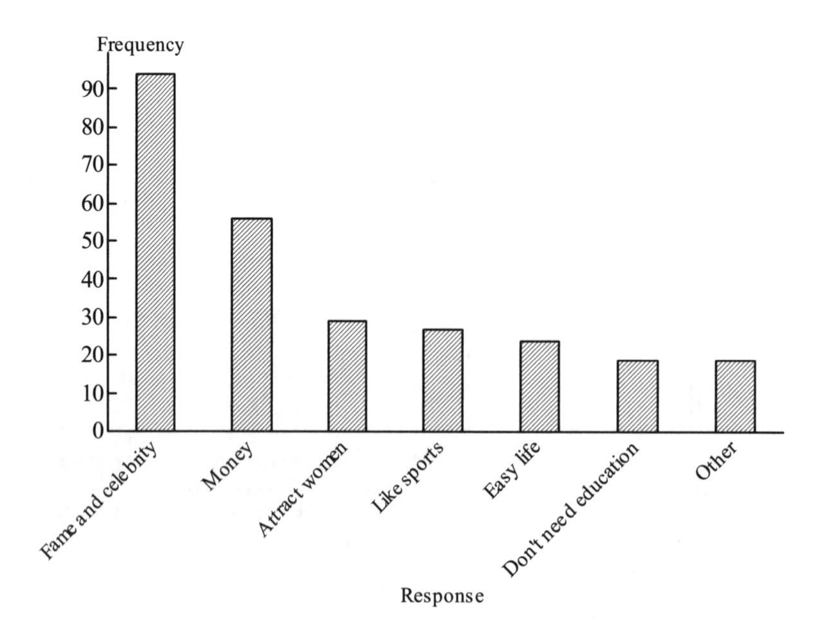

Chapter 2
Collecting Data Sensibly

2.1 **a** This is an observational study.

b No. It is quite possible, for example, that those children who averaged more than two hours of television viewing per day received, generally speaking, a less good education than those who did not, and that it is the less good education, and not the television viewing, that caused the lower reading scores.

2.3 **a** This is an observational study.

b Yes. Since the researchers looked at a random sample of publically accessible MySpace web profiles posted by 18-year-olds, it is reasonable to generalize the stated conclusion to all 18-year-olds with publically accessible MySpace profiles.

c No, it is not reasonable to generalize the stated conclusion to all 18-year-old MySpace users since no users without publically accessible profiles were included in the study.

d No, it is not reasonable to generalize the stated conclusion to all MySpace users with publically accessible profiles since only 18-year-olds were included in the study.

2.5 We will refer to students who have a high school GPA of at least 3.5 and a combined SAT score of over 1200 as "well qualified" students. It is quite possible that well qualified students who go to "most selective" colleges are, on the whole, naturally better motivated than well qualified students who go to "least selective" colleges. Therefore, if all the well qualified students who were admitted to "least selective" colleges were moved to "most selective" colleges, these students would not necessarily achieve the 89% graduation rate achieved by well qualified students who were admitted to "most selective" colleges.

2.7 We are told that moderate drinkers, as a group, tended to be better educated, wealthier, and more active than nondrinkers. It is therefore quite possible that the observed reduction in the risk of heart disease amongst moderate drinkers is caused by one of these attributes and not by the moderate drinking.

2.9 It is not appropriate to make the conclusion stated since it is quite possible that babies born to mothers with diabetes differ, in some relevant way other than the experience of pain in early life, from babies born to mothers without diabetes. For example, it could be suggested that babies born to mothers with diabetes have, due to the exposure during early development to their mothers' blood, a greater susceptibility to stress in general than babies born to mothers without diabetes. Thus the grimacing and crying observed amongst these babies when having blood drawn could be caused by this greater susceptibility to stress and not by the pain experienced early in life.

2.11 **a** The data would need to be collected from a simple random sample of affluent Americans.

b No. Since the survey included only affluent Americans the result cannot be generalized to all Americans.

2.13 Method 1: Using a computer list of the graduates, number the graduates 1-140. Use a random number generator on a calculator or computer to randomly select a whole number between 1 and 140. The number selected represents the first graduate to be included in the sample. Repeat the number selection, ignoring repeated numbers, until 20 graduates have been selected.
Method 2: Using a computer list of the graduates, number the graduates 001-140. Take the first three digits from the left hand end of a row from a table of random digits. If the three-digit number formed is between 001 and 140 inclusive, the graduate with that number should be the first graduate in the sample. If the number formed is not between 001 and 140 inclusive, the number should be ignored. Repeat the process described for the next three digits in the random number table, and continue in the same way until 20 graduates have been selected. (Three-digit numbers that are repeats of numbers previously selected should be ignored.)

2.15 Using a computer list of the cases, number the cases 1-870. Use a random number generator on a calculator or computer to randomly select a whole number between 1 and 140. The number selected represents the first case to be included in the sample. Repeat the number selection, ignoring repeated numbers, until 50 cases have been selected.

2.17 The method used by researcher B is preferable. It is quite possible that the rows will differ in terms of the sugar content of fruit from the trees. In the method used by researcher A the sample obtained will necessarily include trees from exactly six rows. However, in the method used by researcher B the sample will very likely include trees from a much greater number of rows, and is therefore more likely to be representative of the population of trees.

2.19 **a** Using the list, first number the part time students 1-3000. Use a random number generator on a calculator or computer to randomly select a whole number between 1 and 3000. The number selected represents the first part time student to be included in the sample. Repeat the number selection, ignoring repeated numbers, until 10 part time students have been selected. Then number the full time students 1-3500 and select 10 full time students using the same procedure.

b No. With 10 part time students being selected out of a total of 3000 part time students, the probability of any particular part time student being selected is 10/3000 = 1/300. Applying a similar argument to the full time students, the probability of any particular full time student being selected is 10/3500 = 1/350. Since these probabilities are different, it is not the case that every student has the same chance of being included in the sample.

2.21 **a** The pages of the book have already been numbered between 1 and the highest page number in the book. Use a random number generator on a calculator or computer to randomly select a whole number between 1 and the highest page number in the book. The number selected will be the first page to be included in the sample. Repeat the number selection, ignoring repeated numbers, until the required number of pages has been selected.

b Pages that include exercises tend to contain more words than pages that do not include exercises. Therefore, it would be sensible to stratify according to this criterion. Assuming that 20 non-exercise pages and 20 exercise pages will be included in the sample, the sample should be selected as follows. Use a random number generator to randomly select a whole number between 1 and the highest page number in the book. The number selected will be the first page to be included in the sample. Repeat the number selection, ignoring repeated numbers and keeping track of the number of pages of each type selected, until 20 pages of one type have been selected. Then continue in the same way, but ignore numbers corresponding to pages of that type. When 20 pages of the other type have been selected, stop the process.

c Randomly select one page from the first 20 pages in the book. Include in your sample that page and every 20th page from that page onwards.

d Roughly speaking, in terms of the numbers of words per page, each chapter is representative of the book as a whole. It is therefore sensible for the chapters to be used as clusters. Using a random number generator randomly choose three chapters. Then count the number of words on each page in those three chapters.

e Answers will vary.

f Answers will vary.

2.23 The researchers should be concerned about nonresponse bias. Only a small proportion (20.7%) of the selected households completed the interview, and it is quite possible that those households who did complete the interview are different in some relevant way concerning Internet use from those who did not.

2.25 First, the participants in the study were all students in an upper-division communications course at one particular university. It is not reasonable to consider these students to be representative of all students with regard to their truthfulness in the various forms of communication. Second, the students knew during the week's activity that they were surveying themselves as to the truthfulness of their interactions. This could easily have changed their behavior in particular social contexts and therefore could have distorted the results of the study.

2.27 First, the people who responded to the print and online advertisements might be different in some way relevant to the study from the population of people who have online dating profiles. Second, only the *Village Voice* and Craigslist New York City were used for the recruitment. It is quite possible that people who read that newspaper or access those websites differ from the population in some relevant way, particularly considering that they are both New York City based publications.

2.29 The individuals within each stratum should on the whole be similar in terms of the topic of the study. This is true of the proposed strata in Scheme 2, since it is likely that college students will on the whole be similar in their opinions regarding the possible tax increase; likewise nonstudents who work full time will on the whole be similar in their opinions regarding the possible tax increase, and nonstudents who do not work full time will on the whole be similar in their opinions regarding the possible tax increase. Scheme 1, however, is not suitable since we have no reason to believe that people within the proposed first-letter-of-last-name strata will be similar in terms of their attitudes to the possible tax increase. Similarly the suggested stratification in Scheme 3 is very unlikely to produce homogeneous groups.

2.31 Different subsets of the population might have responded by different methods. For example, it is quite possible that younger people (who might generally be in favor of continuing the parade) chose to respond via the Internet while older people (who might on the whole be against the parade) chose to use the telephone to make their responses.

2.33 **a** Binding strength

b Type of glue

c The extraneous variables mentioned are the number of pages in the book and whether the book is bound as a hardback or a paperback. Further extraneous variables that might be considered include the weight of the material used for the cover and the type of paper used.

2.35 Random assignment should have been used to determine, for each cyclist, which drink would be consumed during which break.

2.37 We rely on random assignment to produce comparable experimental groups. If the researchers had hand-picked the treatment groups, they might unconsciously have favored one group over the other in terms of some variable that affects the subjects' ability to deal with multiple inputs.

2.39 **a** If the participants had been able to choose their own avatars, then it is quite possible, for example, that people with a lot of self confidence would tend to choose the attractive avatar while those with less self confidence would tend to choose the unattractive avatar. Then, if the same result was obtained as the one described in the paper, it would be impossible to tell whether the greater closeness achieved by those with the attractive avatar came about as a result of the avatar or as a result of those people's greater self confidence.

b

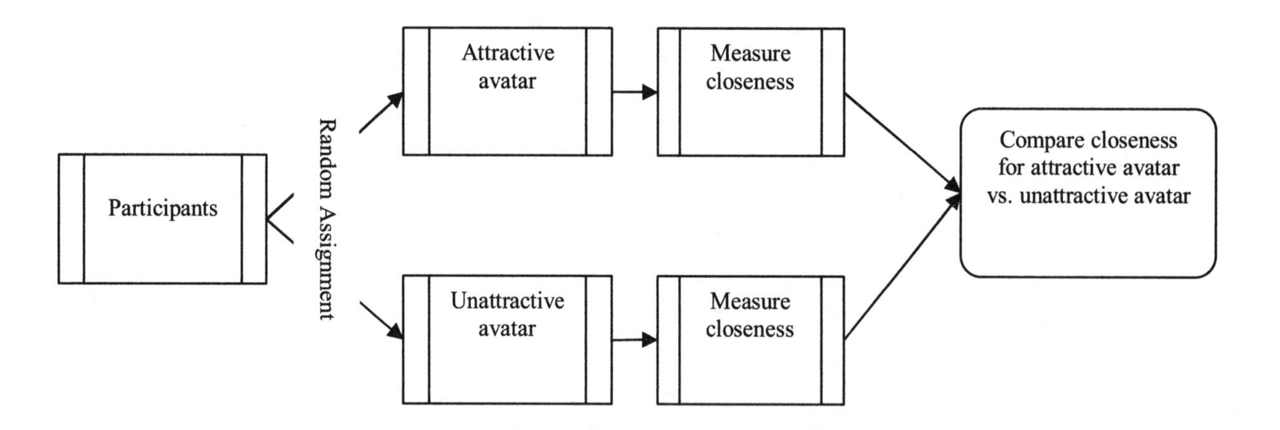

2.41 We rely on random assignment to produce comparable experimental groups. If the researchers had hand-picked the treatment groups, they might unconsciously have favored one group over the other in terms of some variable that affects the subjects' ability to learn through video gaming activity.

2.43

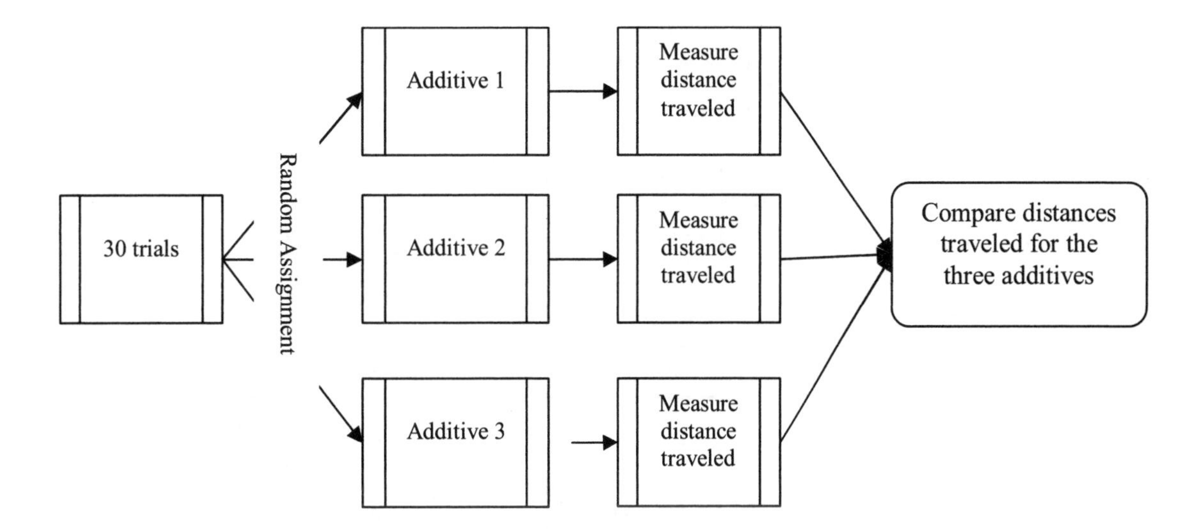

2.45 **a** The improvement in group 3 compared to group 1 cannot be attributed to the use of Sweet Talk since group 3 differs from group 1 in two respects: the incorporation of Sweet Talk and the use of the new intensive insulin therapy in place of the conventional insulin therapy. Therefore it is not possible to tell whether the improvement is attributable to Sweet Talk, the intensive insulin therapy, or a combination of the two. (Note that the fact that there is no significant difference in the results for groups 1 and 2 suggests that Sweet Talk is not beneficial when used in conjunction with the conventional insulin treatment. It does not tell us whether Sweet Talk would be helpful when the intensive insulin treatment is being used.)

b The experiment needs to be modified by the addition of a group (group 4) that receives the intensive insulin therapy without Sweet Talk support. Then a comparison between the results of groups 3 and 4 will tell the experimenters whether Sweet Talk presents an improvement when the intensive insulin therapy is being used.

c

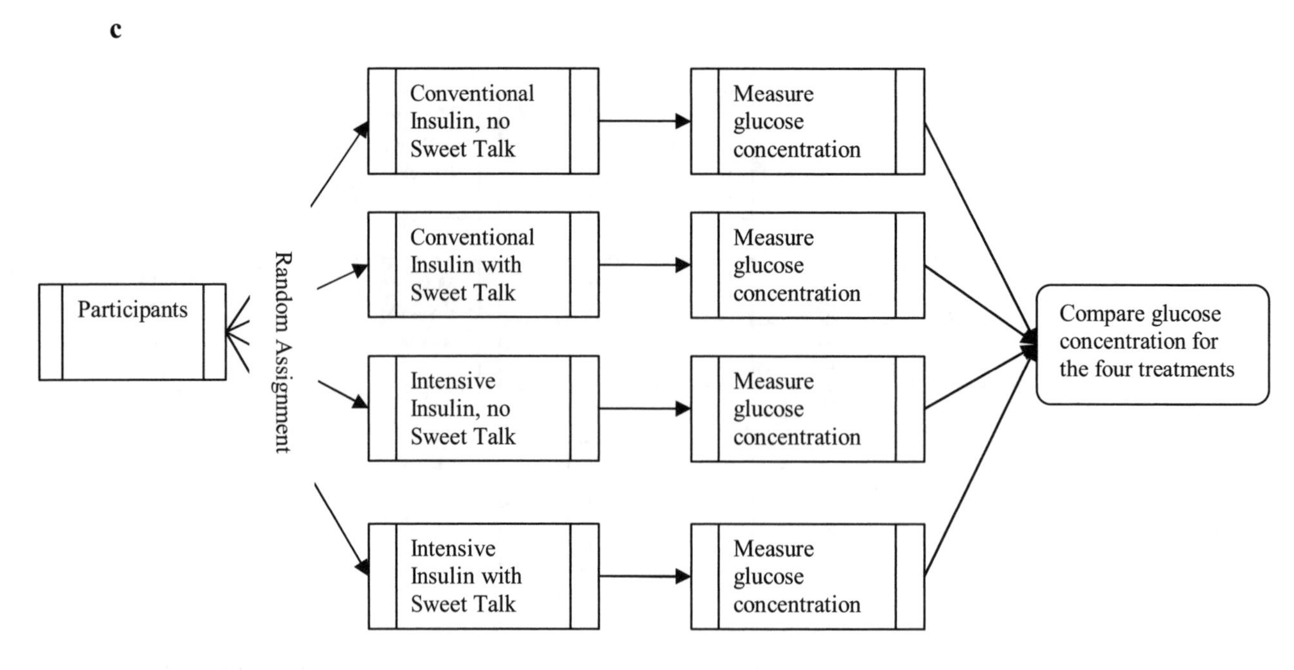

2.47 **a** Red wine, yellow onions, black tea

b Absorption of flavonol into the blood

c Gender, amount of flavonols consumed apart from experimental treatment, tolerance of alcohol in wine

2.49 "Blinding" is ensuring that the experimental subjects do not know which treatment they were given and/or ensuring that the people who measure the response variable do not know who was given which treatment. When this is possible to implement, it is useful that the subjects do not know which treatments they were given since, if a person knows what treatment he/she was given, this knowledge could influence the person's perception of the response variable, or even, through psychological processes, have a direct effect on the response variable. If the response variable is to be measured by a person other than the experimental subjects it is useful if this person doesn't know who received which treatment since, if this person *does* know who received which treatment, then this could influence the person's perception of the response variable.

2.51 **a** In order to know that the results of this experiment are valid it is necessary to know that the assignment of the women to the groups was done randomly. For suppose, for example, that the women were allowed to choose which groups they went into. Then it would be quite possible, for instance, that women who are particularly social by nature, and therefore whose health would be enhanced by any regular social gathering, would choose the more interesting sounding art discussions, while those less social by nature (and therefore less likely to be helped by social gatherings) would choose the more conventional discussions of hobbies and interests. Then it would be impossible to tell whether the stated results were caused by the discussions of art or by the greater social nature of the women in the art discussion group.

b Suppose that all the women took part in weekly discussions of art, and that generally an improvement in the medical conditions mentioned was observed amongst the subjects. Then it would be impossible to tell whether these health improvements had been caused by the discussions of art or by some factor that was affecting all the subjects, such as an improvement in the weather over the four months. By including a control group, and by observing that the improvements did not take place (generally speaking) for those in the control group, factors such as this can be discounted, and the discussions of art are established as the cause of the improvements.

2.53 We will assume that only four colors will be compared, and that only headache sufferers will be included in the study.

Prepare a supply of "Regular Strength" Tylenol in four different colors: white (the current color of the medication, and therefore the "control"), red, green, and blue. Recruit 20 volunteers who suffer from headaches. Instruct each volunteer not to take any pain relief medication for a week. After that week is over, issue each volunteer a supply of all four colors. Give each volunteer an order in which to use the colors (this order would be determined randomly for each volunteer). Instruct the volunteers to use one fixed dose of the medication for each headache over a period of four weeks, and to note on a form the color used and the pain relief achieved (on a scale of 0-10, where 0 is no pain relief and 10 is complete pain relief). At the end of the four weeks gather the results and compare the pain relief achieved by the four colors.

2.55 Suppose that the dog handlers and/or the experimental observers had known which patients did and did not have cancer. It would then be possible for some sort of (conscious or unconscious) communication to take place between these people and the dogs so that the dogs would pick up the conditions of the patients from these people rather than through their perception of the patients' breath. By making sure that the dog handlers and the experimental observers do not know who has the disease and who does not it is ensured that the dogs are getting the information from the patients.

2.57 **a** If the judges had known which chowder came from which restaurant then it is unlikely that Denny's chowder would have won the contest, since the judges would probably be conditioned by this knowledge to choose chowders from more expensive restaurants.

b In experiments, if the people measuring the response are not blinded they will often be conditioned to see different responses to some treatments over other treatments, in the same way as the judges would have been conditioned to favor the expensive restaurant chowders. It is therefore necessary that the people measuring the response should not know which subject received which treatment, so that the treatments can be compared on their own merits.

2.59 **a** A placebo group would be necessary if the mere thought of having amalgam fillings could produce kidney disorders. However, since the experimental subjects were sheep the researchers do not need to be concerned that this would happen.

b A resin filling treatment group would be necessary in order to provide evidence that it is the material in the amalgam fillings, rather than the process of filling the teeth, or just the presence of foreign bodies in the teeth, that is the cause of the kidney disorders. If the amalgam filling group developed the kidney disorders and the resin filling group did not, then this would provide evidence that it is some ingredient in the amalgam fillings that is causing the kidney problems.

c Since there is concern about the effect of amalgam fillings it would be considered unethical to use humans in the experiment.

2.61 Answers will vary.

2.63 Answers will vary.

2.65 Answers will vary.

2.67 **a** This is an observational study.

b In order to evaluate the study, we need to know whether the sample was a random sample.

c No. Since the sample used in the Healthy Steps study was known to be nationally representative, and since the paper states that, compared with the HS trial, parents in the study sample were disproportionately older, white, more educated, and married, it is clear that it is not reasonable to regard the sample as representative of parents of all children at age 5.5 years.

d The potential confounding variable mentioned is what the children watched.

e The quotation from Kamila Mistry makes a statement about cause and effect and therefore is inconsistent with the statement that the study can't show that TV was the cause of later problems.

2.69 Answers will vary.

2.71 The first criticism describes measurement bias. Asking people whether they are talking less to family and friends on the phone could be a biased measure of increased social isolation. First, people might be reluctant to give truthful answers to this question, and second, the question ignores face-to-face contact with family and friends. It is possible, for example, that face-to-face interaction might be increasing while phone contact is decreasing. The second criticism describes selection bias. Since this survey about Internet use was based on a group of people who were induced to participate by the offer of free Internet service, it is not reasonable generalize the results to all US adults.

2.73 We rely on random assignment to produce comparable experimental groups. If the researchers had hand-picked the treatment groups, they might unconsciously have favored one group over the other in terms of some variable that affects the ability of the people at the centers to respond to the materials provided.

2.75 **a** Observational study

b It is quite possible that the children who watched large amounts of TV in their early years were also those, generally speaking, who received less attention from their parents, and it was the lack of attention from their parents that caused the later attention problems, not the TV-watching.

2.77 It is possible, for example, that people who are not married are more likely to go out alone (except for the widowed, who are older and therefore tend to stay home). It could then be this going out alone that is causing the risk of being a victim of violent crime, not the marital status.

2.79 All the participants were women, from Texas, and volunteers. All three of these facts tell us that it is likely to be unreasonable to generalize the results of the study to all college students.

2.81 **a** The extraneous variables identified are gender, age, weight, lean body mass, and capacity to lift weights. They were dealt with by direct control: all the volunteers were male, about the same age, and similar in weight, lean body mass, and capacity to lift weights.

b Yes, it is important that the men were not told which treatment they were receiving, otherwise the effect of giving a placebo would have been removed. If the participants *were* told which treatment they were receiving, then those taking the creatine would have the additional effect of the mere taking of a supplement thought to be helpful (the placebo effect) and those getting the fake preparation would not get this effect. It would then be impossible to distinguish the influence of the placebo effect from the effect of the creatine itself.

c Yes, it would have been useful if those measuring the increase in muscle mass had not know who received which treatment. It is possible that, through having this knowledge, the people would have been unconsciously influenced into exaggerating the increase in muscle mass for those who took the creatine.

2.83 The design could be completely randomized or could involve blocking. The following is a completely randomized design.

Divide the plot into a 4 by 4 grid consisting of 16 equally sized square subplots. Number the subplots 1-16. Use a random number generator to select integers between 1 and 16 inclusive. Ignoring repeats, the subplots represented by the first four integers will receive undisturbed native grasses. The subplots represented by the following four integers will receive managed native grasses. The subplots represented by the following four integers will receive undisturbed nonnative grasses. The remaining four subplots will receive managed nonnative grasses.

Some possible confounding variables are the amount of light a subplot receives, the amount of moisture in a subplot, and whether or not a subplot is on the boundary of the grid. (One of these variables, amount of light, for example, will actually *be* a confounding variable if one particular type of grass is assigned to subplots with more light than the other types of grass.)

(A design using blocking would need to include blocks consisting of subplots that are similar in terms of one or more possible confounding variables. The subplots within each block would be randomly assigned to the four grasses.)

This is an experiment, since the treatments (the different types of grass) are imposed on the subplots, rather than using areas of land that already have the types of grass mentioned.

2.85 There are many possible designs. We will describe here a design that blocks for the day of the week and the section of the newspaper in which the advertisement appears. For the sake of argument we will assume that the mortgage lender is interested in advertising on only two days of the week (Monday and Tuesday) and that there are three sections in the newspaper (A, B, and C). We will refer to the three types of advertisement as Ad 1, Ad 1, and Ad 3.

The experimental units are 18 issues of the newspaper (that is, 18 dates) consisting of Mondays and Tuesdays over 9 weeks. Use a random process to decide which three Mondays will receive advertisements in Section A, which three Mondays will receive advertisements in Section B, and which three Mondays will receive advertisements in Section C. Do the same for the nine Tuesdays. We have now effectively split the 18 issues into the six blocks shown below. (There are 3 issues in each block.)

Mon, Sect A	Mon, Sect B	Mon, Sect C
Tue, Sect A	Tue, Sect B	Tue, Sect C

Now randomly assign the three issues in each block to the three advertisements. (Ad 1 is then appearing on three Mondays, once in each section, and on three Tuesdays, once in each section. The same applies to Ad 2 and Ad 3.) The response levels for the three advertisements can now be compared (as can the three different sections and the two different days).

Chapter 3
Graphical Methods for Describing Data

3.1

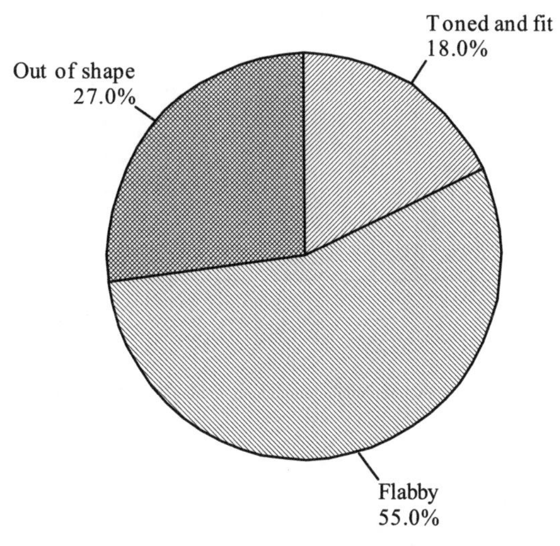

3.3 **a** The second and third categories ("Permitted for business purposes only" and "Permitted for limited personal use" were combined into one category ("No, but some limits apply").

b

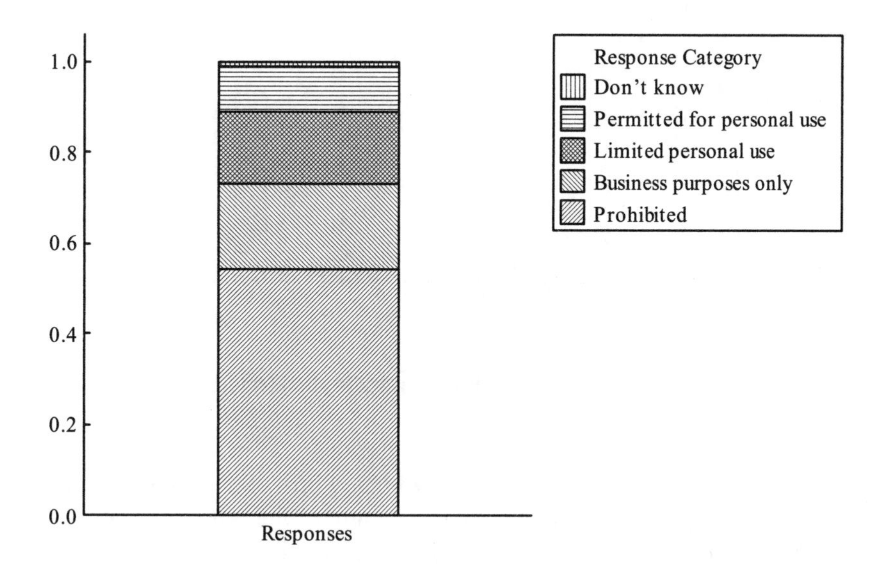

c Pie chart, regular bar graph

3.5

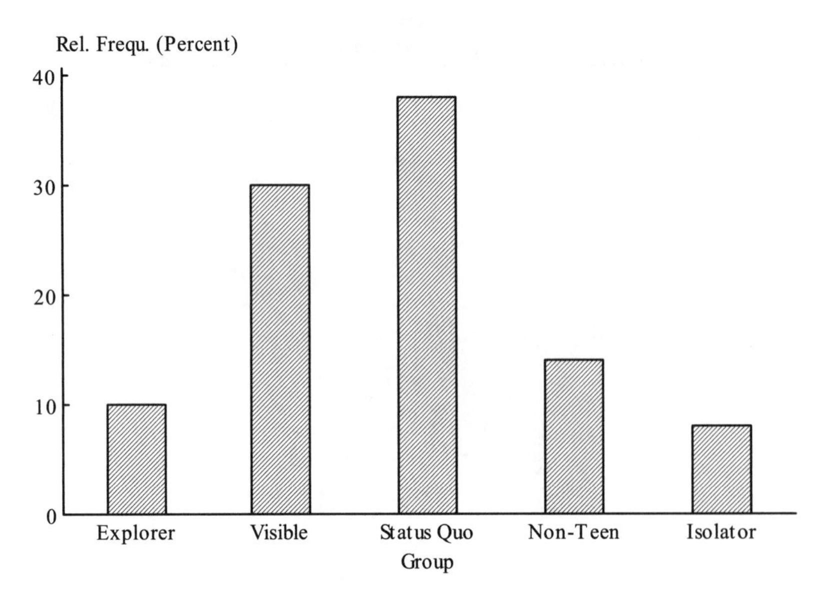

Since the number of categories is relatively high, a bar graph is suitable.

3.7 **a**

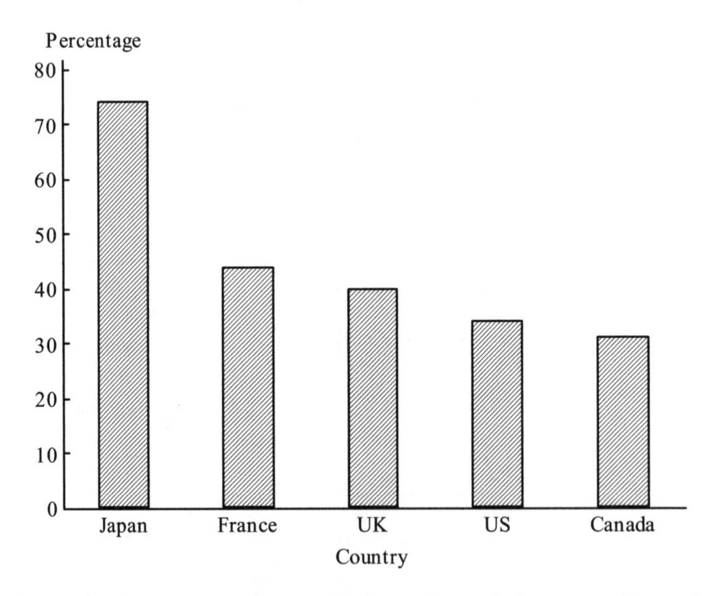

b Were the surveys carried out on random samples of married women from those countries? How were the questions worded?

c In one country, Japan, the percentage of women who say they never get help from their husbands is far higher than the percentages in any of the other four countries included. The percentages in the other four countries are similar, with Canada showing the lowest percentage of women who say they do not get help from their husbands.

3.9

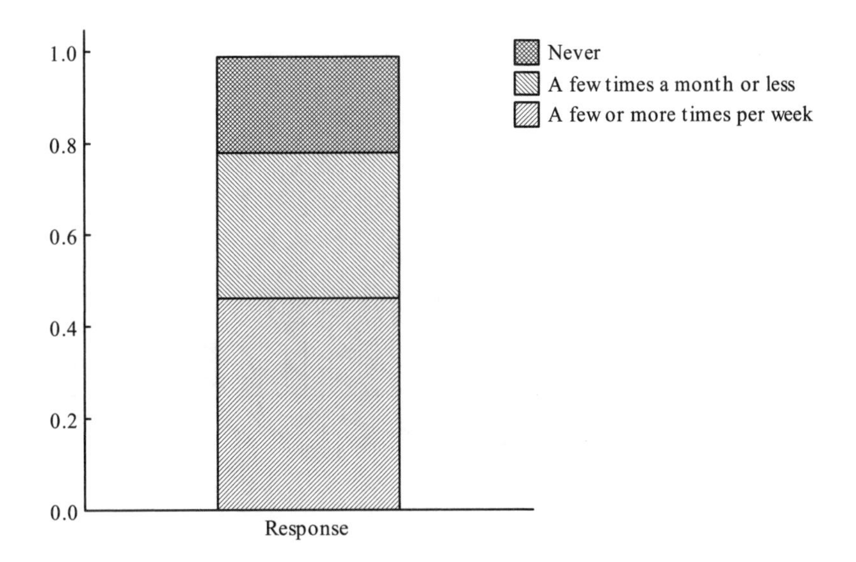

3.11 **a**

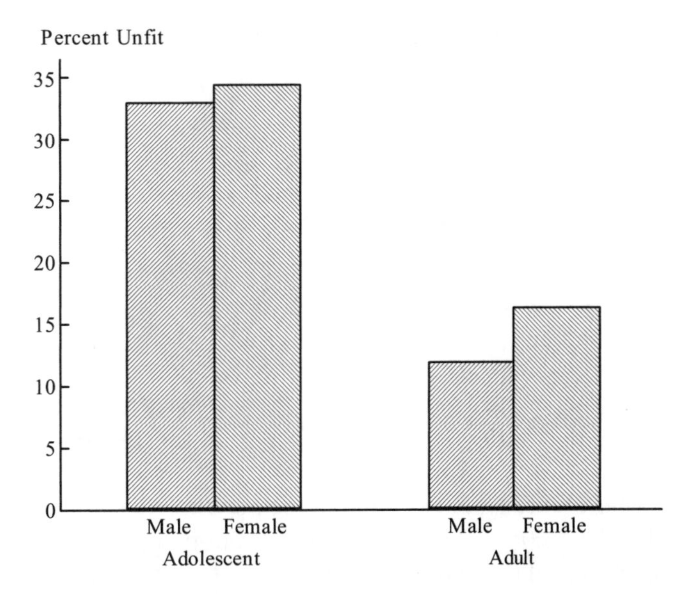

b The comparative bar graph shows that a much higher proportion of adolescents are unfit than adults. It also shows that while amongst adolescents the rates of unfitness are roughly the same for females and males, amongst adults the rate is significantly higher for females than it is for males.

3.13 **a** No. A pie chart is unsuitable when there is such a large number of categories.

b

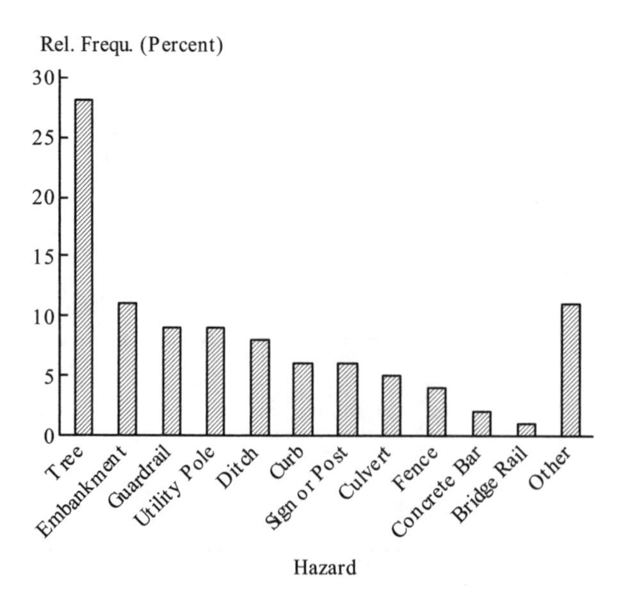

Yes, it is easier to see the differences between the relative frequencies for the different hazards, particularly for those with small relative frequencies.

3.15

10	578
11	79
12	1114
13	001122478899
14	0011112235669
15	11122445599
16	1227
17	1
18	
19	
20	8

Stem: Ones
Leaf: Tenths

A typical number of births per thousand of the population is around 14, with most birth rates concentrated in the 13.0 to 15.9 range. The distribution has just one peak (at the 14-15 class). There is an extreme value, 20.8, at the high end of the data set, and this is the only birth rate above 17.1. The distribution is not symmetrical, since it has a greater spread to the right of its center than to the left.

3.17 **a**

0H	55567889999
1L	0000111113334
1H	556666666667789
2L	00001122233
2H	5

Stem: Tens
Leaf: Ones

A typical percentage of households with only a wireless phone is around 15.

b

West		East
998	0H	555789
110	1L	00011134
8766	1H	666
21	2L	00
5	2H	

Stem: Tens
Leaf: Ones

A typical percentage of households with only a wireless phone for the West is around 16, which is greater than for the East (around 11). There is a slightly greater spread of values in the West than in the East, with values in the West ranging from 8 to 25 (a range of 17) and values in the East ranging from 5 to 20 (a range of 15). The distribution for the West is roughly symmetrical, while the distribution in the East shows a slightly greater spread to the right of its center than to the left. Neither distribution has any outliers.

3.19 **a**

Stem	Leaves
−1	100
−0	9999888877655555544443322221110
0	000011244577
1	179
2	2

Stem: Tens
Leaf: Ones

b Split each stem into two, one taking the lower leaves (0-4) and the other taking the higher leaves (5-9). So, for example, the stem "0" would be split into "0L" and "0H", with 0L taking the leaves "000011244" and 0H taking the leaves "44577".

c The three states with the greatest percentage increase in the number of 25- to 44-year-olds are Nevada, Utah, and Arizona, all desert states.

3.21

Stem	Leaves
0t	333
0f	44444455555
0s	666666666677777777777
0*	88888888999
1.	0000

Stem: Tens
Leaf: Ones

The stem-and-leaf display shows that the distribution of high school dropout rates is roughly symmetrical. A typical dropout rate is 7%. The great majority of rates are between 4% and 9%, inclusive.

3.23

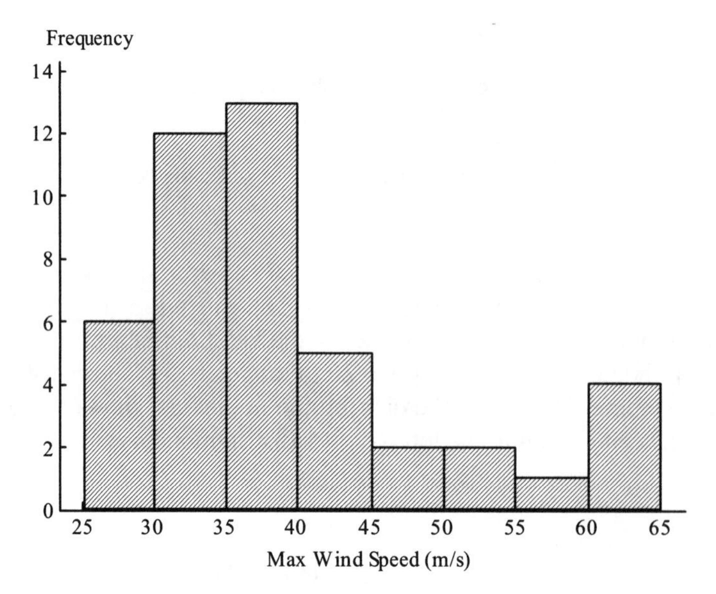

The distribution of maximum wind speeds is positively skewed and is bimodal, with peaks at the 35-40 and 60-65 intervals.

3.25 **a**

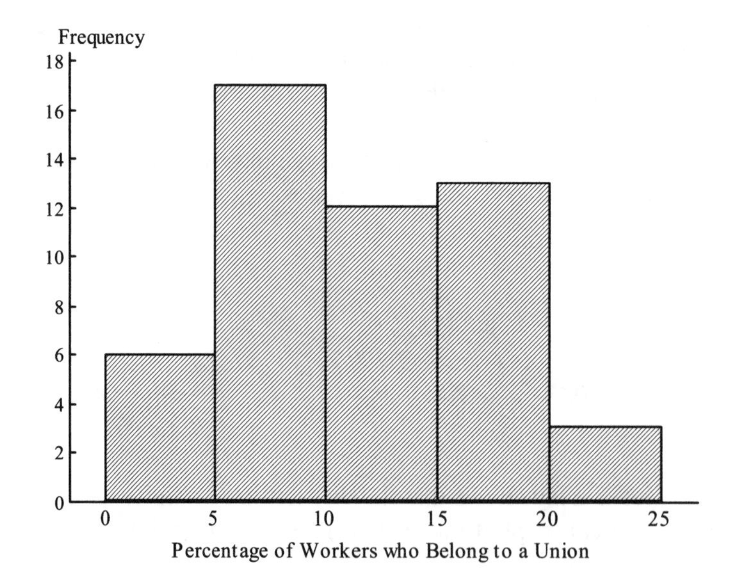

b

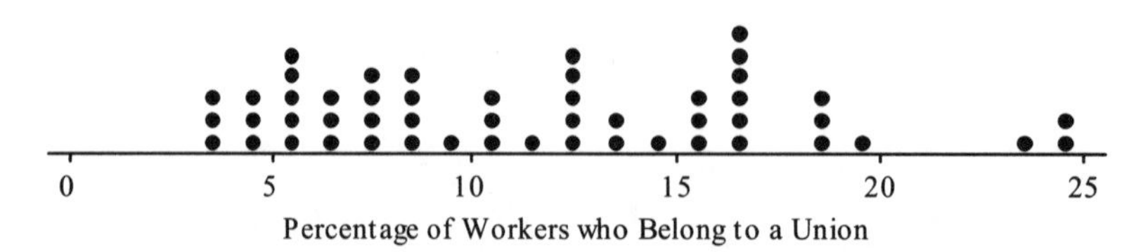

A typical percentage of workers belonging to a union is around 11, with values ranging from 3.5 to 24.9. There are three states whose percentages stand out as being higher than those of the rest of the states. The distribution is positively skewed.

c The dotplot is more informative as it shows where the data points actually lie. For example, in the histogram we can tell that there are 3 observations in the 20 to 25 interval, but we don't see the actual values and miss the fact that these values are actually considerably higher than the other values in the data set.

d

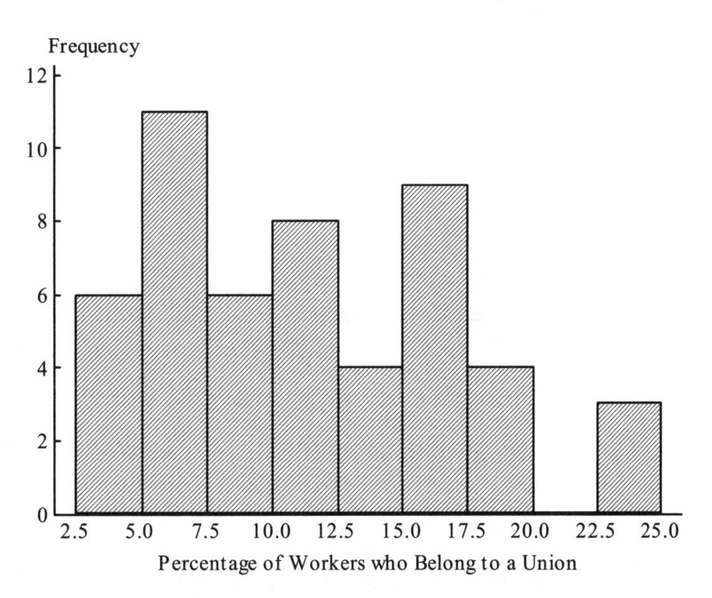

The histogram in part (a) could be taken to imply that there are states with a percent of workers belonging to a union near zero. It is clear from this second histogram that this is not the case. Also, the second histogram shows that there is a gap at the high end and that the three largest values are noticeably higher than those of the other states. This fact is not clear from the histogram in part (a).

3.27 **a**

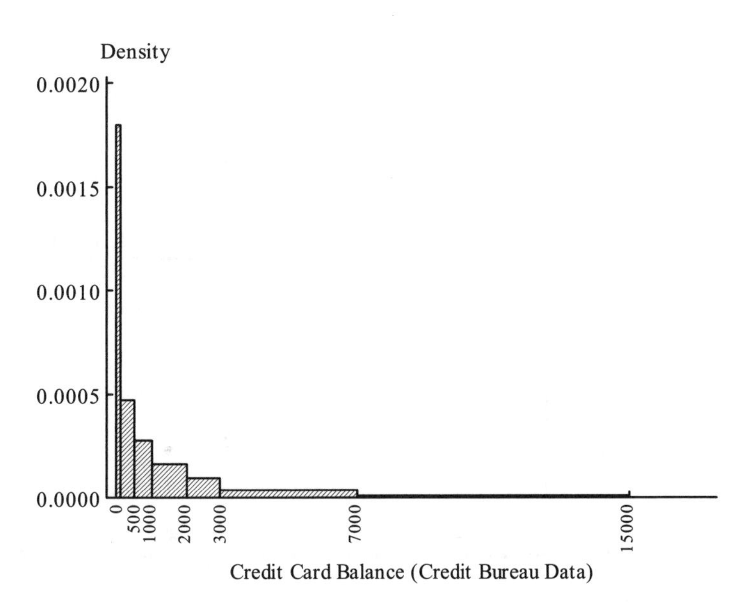

b

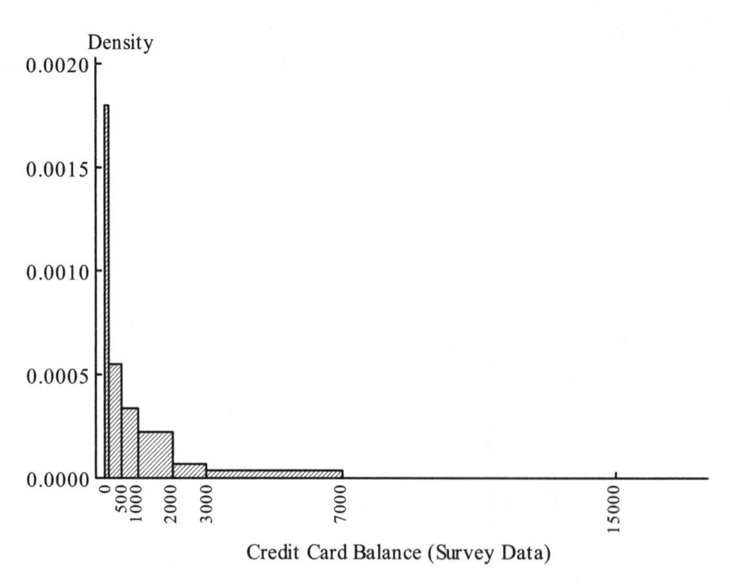

c The histograms are very similar, except that the Credit Bureau results show 7% of students having a debt of $7000 or more, whereas in the survey no student admitted to having a debt this size.

d Yes. It is quite possible that the students who did not respond included those with a debt of over $7000, particularly as students with such a large debt would probably not want to admit it.

3.29 **a** First, the class intervals do not all have the same width, and so use of relative frequency on the y-axis would not be appropriate. Second, we are not given an upper boundary for the last class interval, so we don't have enough information to draw the histogram.

b

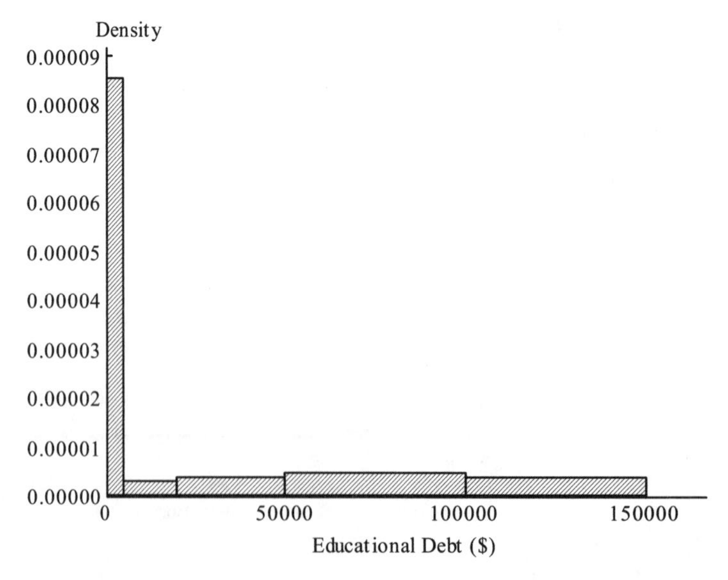

c By far the highest density of educational debts occurs in the $0-5000 range, with 43% of the students having debts in this relatively narrow interval. Amongst the remaining 57% of students there seems to be a roughly symmetrical distribution of debts, with the greatest density of debts occurring in the $50,000-100,000 range.

3.31

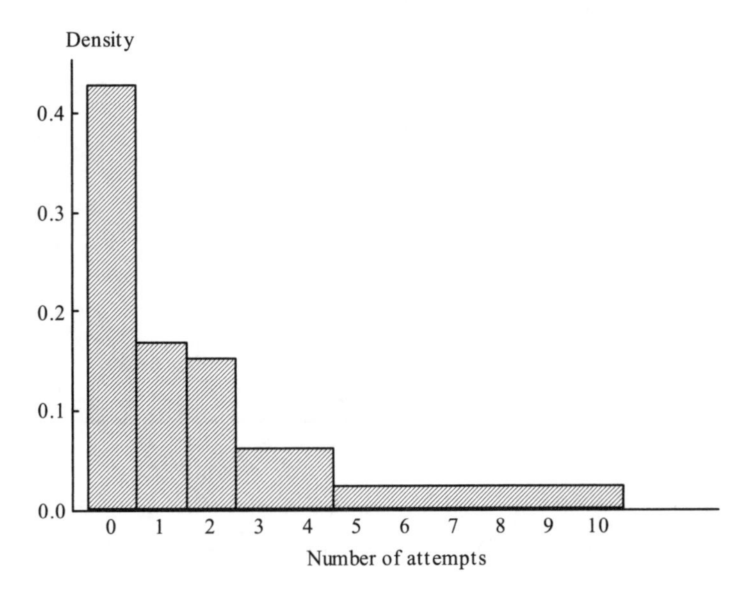

3.33 Answers will vary.

3.35 **a**

Years Survived	Relative Frequency
0 - < 2	.10
2 - < 4	.42
4 - < 6	.02
6 - < 8	.10
8 - < 10	.04
10 - < 12	.02
12 - < 14	.02
14 - < 16	.28

b

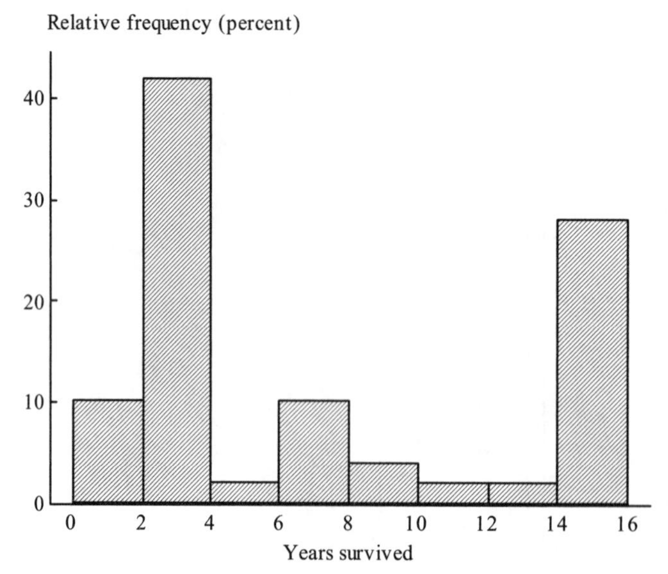

c The histogram shows a bimodal distribution, with peaks at the 2-4 year and 14-16 year intervals. All the other survival times were considerably less common than these two.

d We would need to know that the set of patients used in the study formed a random sample of all patients younger than 50 years old who had been diagnosed with the disease and had received the high dose chemotherapy.

3.37 Answers will vary. One possibility for each part is shown below.

a

Class Interval	100 to <120	120 to <140	140 to <160	160 to <180	180 to <200
Frequency	5	10	40	10	5

b

Class Interval	100 to <120	120 to <140	140 to <160	160 to <180	180 to <200
Frequency	20	10	4	25	11

c

Class Interval	100 to <120	120 to <140	140 to <160	160 to <180	180 to <200
Frequency	33	15	10	7	5

d

Class Interval	100 to <120	120 to <140	140 to <160	160 to <180	180 to <200
Frequency	5	7	10	15	33

3.39

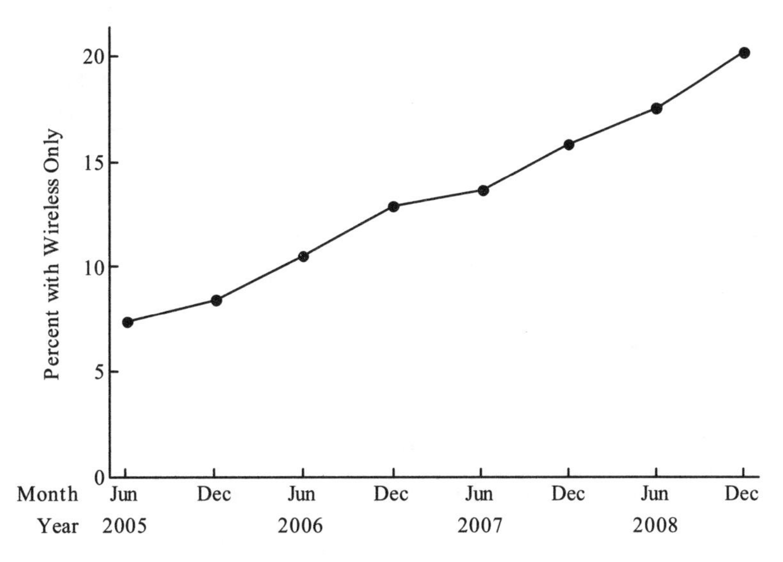

The graph shows an upward trend in the percentage of homes with only a wireless phone service from June 2005 to December 2008. The increase has been at a roughly steady rate, with only the periods June to December 2005 and December 2006 to June 2007 showing a slightly lower rate of growth.

3.41 a

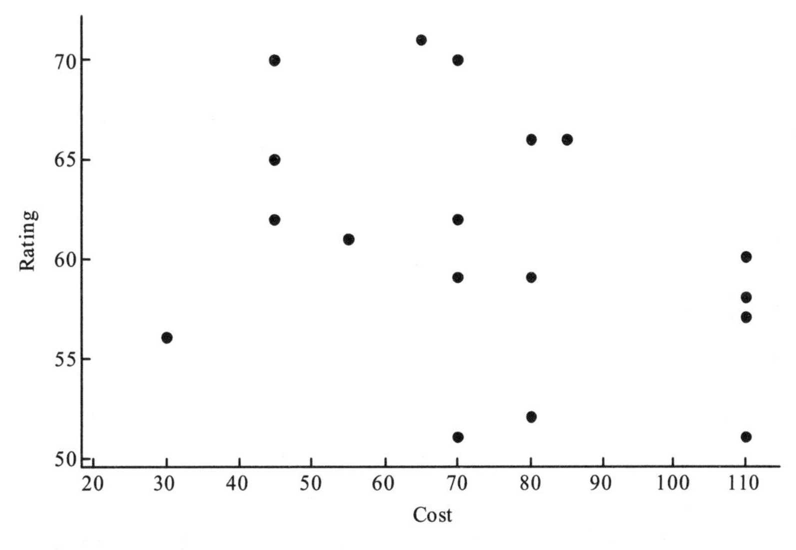

There is a weak relationship between cost and quality rating, with higher costs being loosely associated with lower ratings.

b

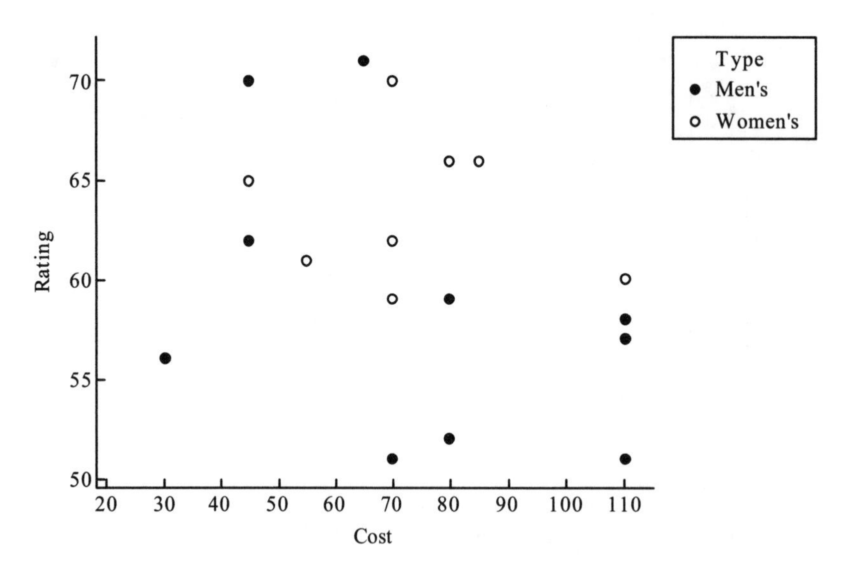

The range of costs for men's athletic shoes is slightly greater than for women's (with just one type of men's shoe providing a cheaper option). For any given cost, there is generally speaking a greater spread of ratings for men's shoes than for women's, with the women's shoes tending to show slightly higher ratings than the men's. For women's shoes the relationship between cost and quality rating is very weak. For men's shoes the relationship is stronger for the women's (and stronger than for the combined data set).

3.43

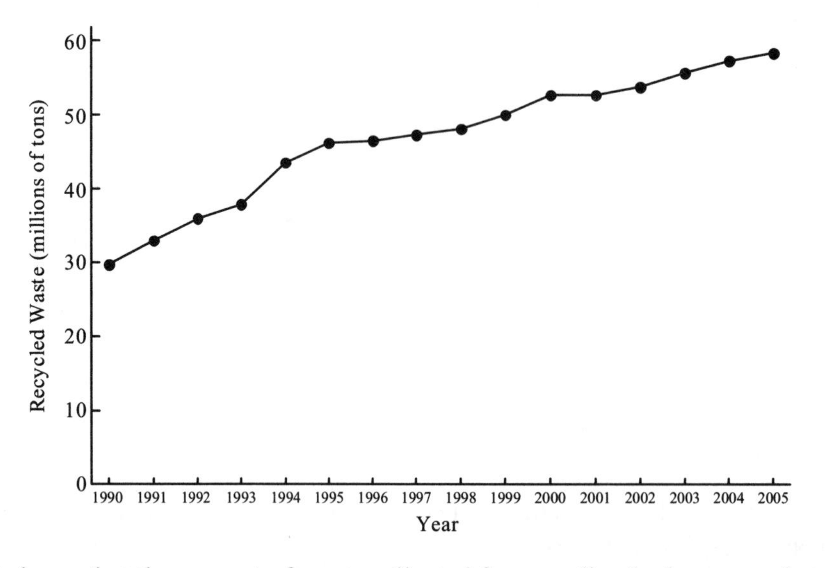

The plot shows that the amount of waste collected for recycling had grown substantially (not slowly, as is stated in the article) in the years 1990 to 2005. The amount increased from under 30 million tons to nearly sixty million tons in that period, which means that the amount had almost doubled.

3.45 According to the 2001 and 2002 data, there are seasonal peaks at weeks 4, 9, and 14, and seasonal lows at weeks 2, 6, 10-12, and 18.

3.47 **a**

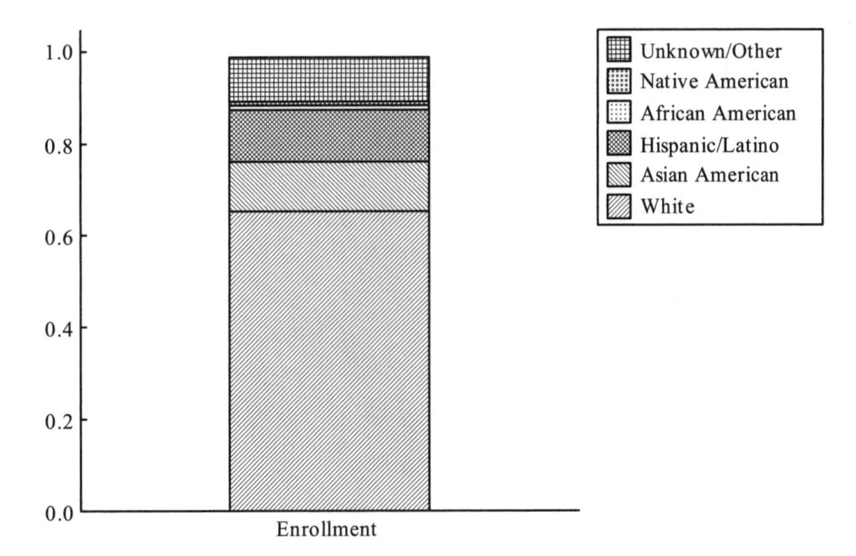

b The graphical display created in Part (a) is more informative, since it gives an accurate representation of the proportions of the ethnic groups.

c The people who designed the original display possibly felt that the four ethnic groups shown in the segmented bar section might seem to be underrepresented at the college if they used a single pie chart.

3.49 The first graphical display is not drawn appropriately. The Z's have been drawn so that their heights are in proportion to the percentages shown. However, the widths and the perceived depths are also in proportion to the percentages, and so neither the areas nor the perceived volumes of the Z's are proportional to the percentages. The graph is therefore misleading to the reader. In the second graphical display, however, *only* the heights of the cars are in proportion to the percentages shown. The widths of the cars are all equal. Therefore the areas of the cars are in proportion to the percentages, and this is an appropriately drawn graphical display.

3.51 The piles of cocaine have been drawn so that their heights are in proportion to the percentages shown. However, the widths are also in proportion to the percentages, and therefore neither the areas (nor the perceived volumes) are in proportion to the percentages. The graph is therefore misleading to the reader.

3.53

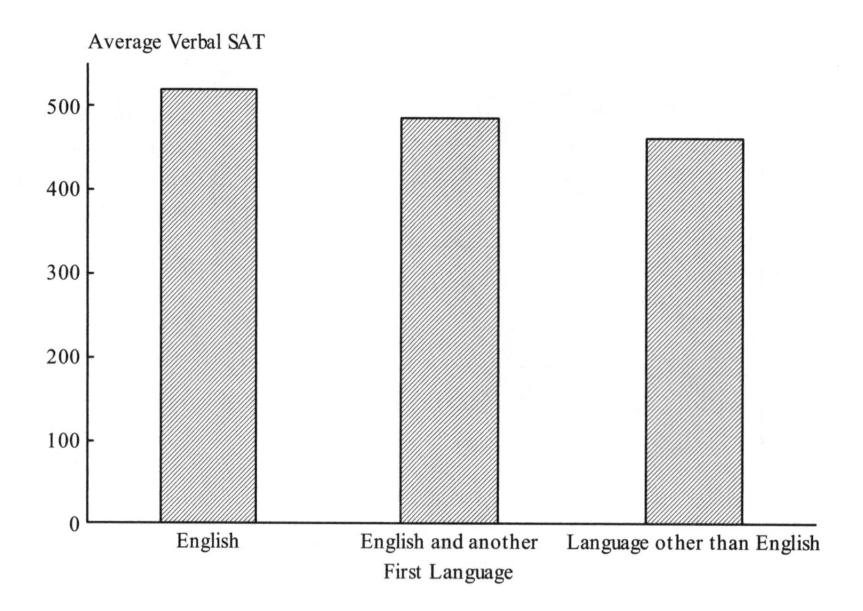

3.55

1	9
2	23788999
3	0011112233459
4	0123

Stem: Tens
Leaf: Ones

A typical calorie content for these light beers is 31 calories per 100 ml, with the great majority lying in the 22-39 range. The distribution is negatively skewed, with one peak (in the 30-39 range). There are no gaps in the data.

3.57 **a**

0	0033344555568888888999999
1	0001223344567
2	001123689
3	0
4	0
5	
6	6

Stem: Tens
Leaf: Ones

b A typical percentage population increase is around 10, with the great majority of states in the 0-29 range. The distribution is positively skewed, with one peak (in the 0-9 range). There are two states showing significantly greater increases than the other 48 states: one at 40 (Arizona) and one at 66 (Nevada).

c

West		East
9988880	0	033344555568889999
432	1	0001234567
982100	2	136
0	3	
0	4	
	5	
6	6	

Stem: Tens
Leaf: Ones

On average, the percentage population increases in the West were greater than those for the East, with a typical value for the West being around 14 and a typical value for the East being around 9. There is a far greater spread in the values in the West, with values ranging from 0 to 66, than in the East where values ranged from 0 to 26. Both distributions are positively skewed, with a single peak for the East data, and two peaks for the West. In the West there are two states showing significantly greater increases than the remaining states, with values at 40 and 60. There are no such extreme values in the East.

3.59 a High graft weight ratios are clearly associated with low body weights (and vice versa), and the relationship is not linear. (In fact there seems to be, roughly speaking, an inverse proportionality between the two variables, apart from a small increase in the graft weight ratios for increasing body weights amongst those recipients with the greater body weights. This is interesting in that an inverse proportionality between the variables would imply that the actual weights of transplanted livers are chosen independently of the recipients' body weights.)

b A likely reason for the negative relationship is that the livers to be transplanted are probably chosen according to whatever happens to be available at the time. Therefore, lighter patients are likely to receive livers that are too large and heavier patients are likely to receive livers that are too small.

3.61 a

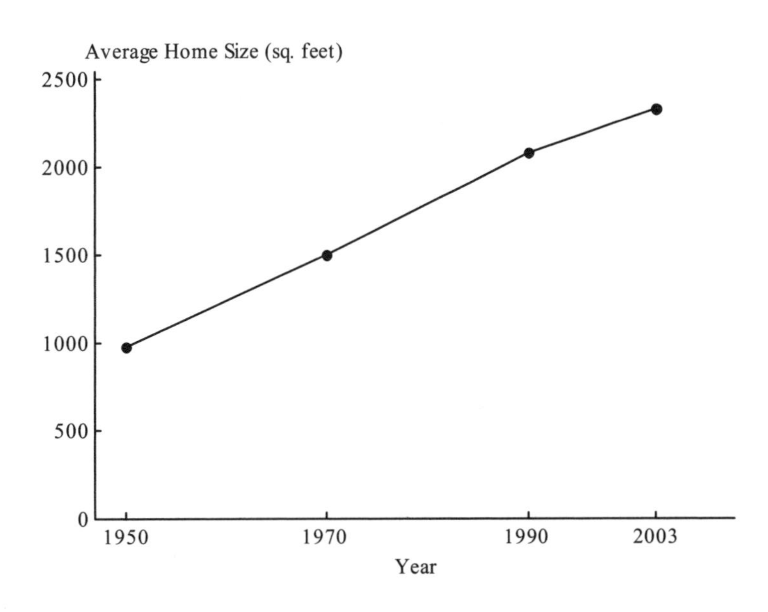

b Continuing the growth trend, we estimate that the average home size in 2010 will be approximately 2500 square feet.

3.63

Disney		Other
975332100	0	0001259
765	1	156
920	2	0
	3	
	4	
4	5	

Stem: Hundreds
Leaf: Tens

On average, the total tobacco exposure times for the Disney movies are higher than the others, with a typical value for Disney being around 90 seconds and a typical value for the other companies being around 50 seconds. Both distributions have one peak and are positively skewed. There is one extreme value (548) in the Disney data, and no extreme value in the data for the other companies. There is a greater spread in the Disney data, with values ranging from 6 seconds to 540 seconds, than for the other companies, where the values range from 1 second to 205 seconds.

3.65 **a**

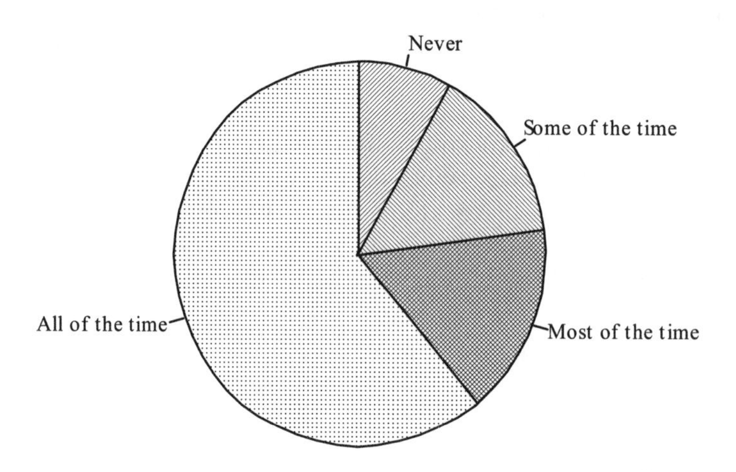

b

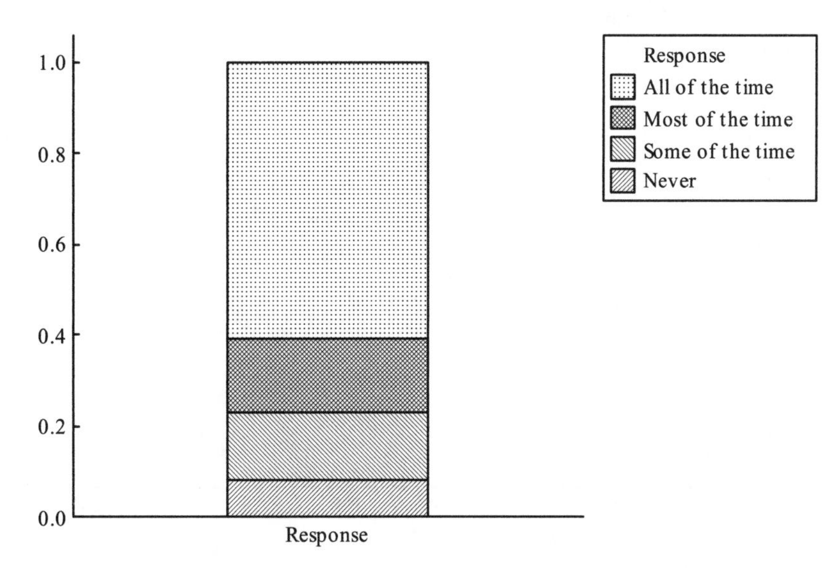

c The segmented bar graph is slightly preferable in that it is a little easier than in the pie chart to see that the proportion of children responding "Most of the time" was slightly higher than the proportion responding "Some of the time."

3.67

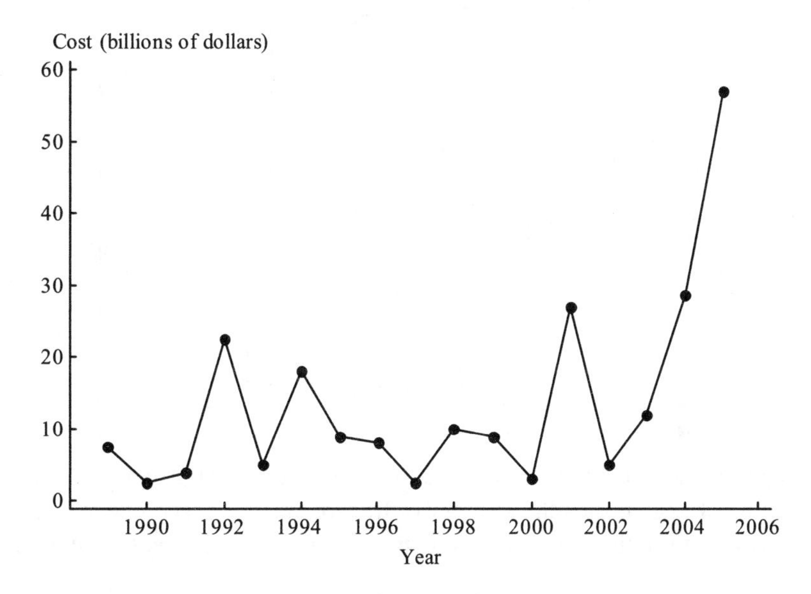

The peaks were probably caused by the incidence of major hurricanes in those years.

3.69 **a**

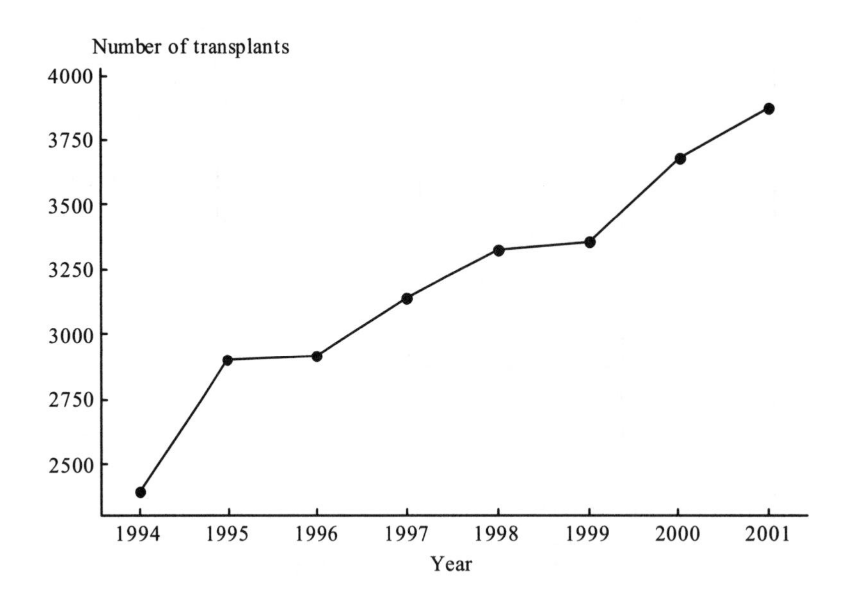

b

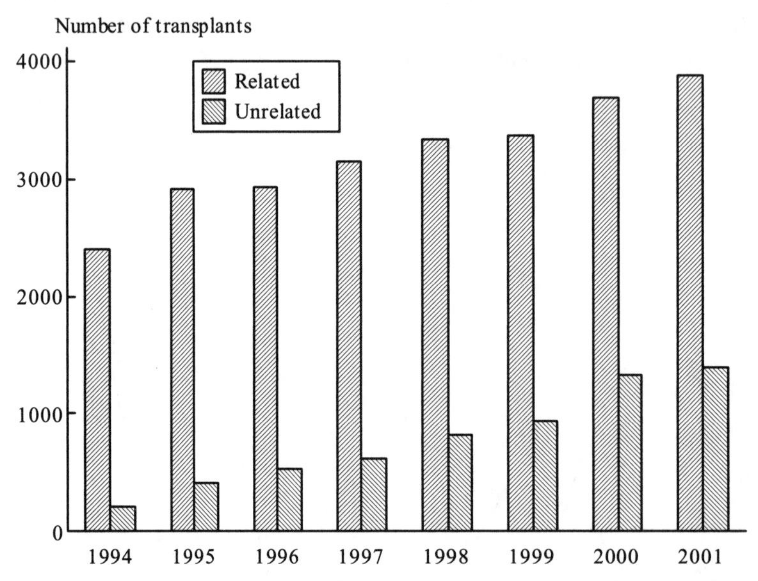

In every year the number of related donors was much greater than the number of unrelated donors. In both categories the number of transplants increased every year, but proportionately the increases in unrelated donors were greater than the increases in related donors.

3.71 a

Skeletal Retention	Frequency
0.15 to <0.20	4
0.20 to <0.25	2
0.25 to <0.30	5
0.30 to <0.35	21
0.35 to <0.40	9
0.40 to <0.45	9
0.45 to <0.50	4
0.50 to <0.55	0
0.55 to <0.60	1

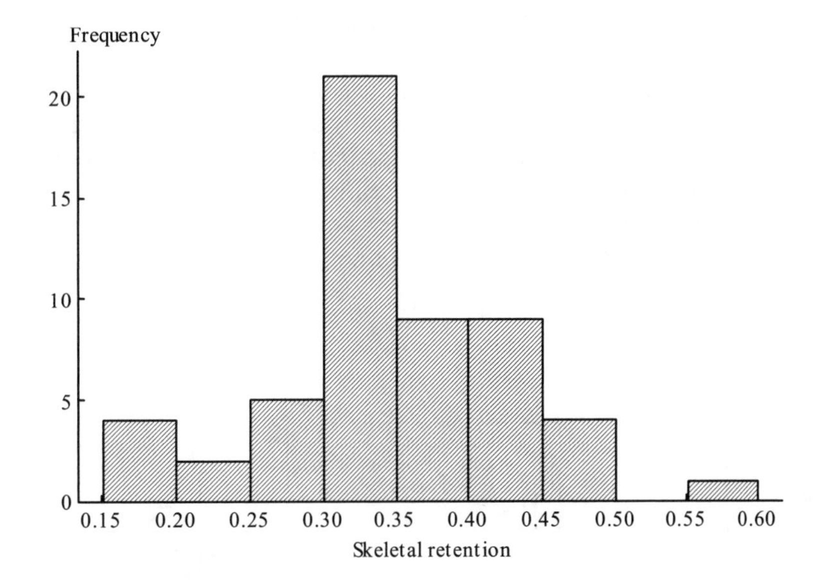

b The histogram is centered at approximately 0.34, with values ranging from 0.15 to 0.5, plus one extreme value in the 0.55-0.6 range. The distribution has a single peak and is slightly positively skewed.

Cumulative Review Exercises

CR3.1 No. It is quite possible, for example, that men who ate a high proportion of cruciferous vegetables generally speaking also had healthier lifestyles than those who did not, and that it was the healthier lifestyles that were causing the lower incidence of prostate cancer, not the eating of cruciferous vegetables.

CR3.3 Very often those who choose to respond generally have a different opinion on the subject of the study from those who do not respond. (In particular, those who respond often have strong feelings against the status quo.) This can lead to results that are not representative of the population that is being studied.

CR3.5 Only a small proportion (around 11%) of the doctors responded, and it is quite possible that those who did respond had different opinions regarding managed care from the majority who did not. Therefore the results could have been very inaccurate for the population of doctors in California.

CR3.7 Suppose, for example, the women had been allowed to choose whether or not they participated in the program. Then it is quite possible that generally speaking those women with more social awareness would have chosen to participate, and those with less social awareness would have chosen not to. Then it would be impossible to tell whether the stated results came about as a result of the program or of the greater social awareness amongst the women who participated. By randomly assigning the women to participate or not, comparable groups of women would have been obtained.

CR3.9 **a**

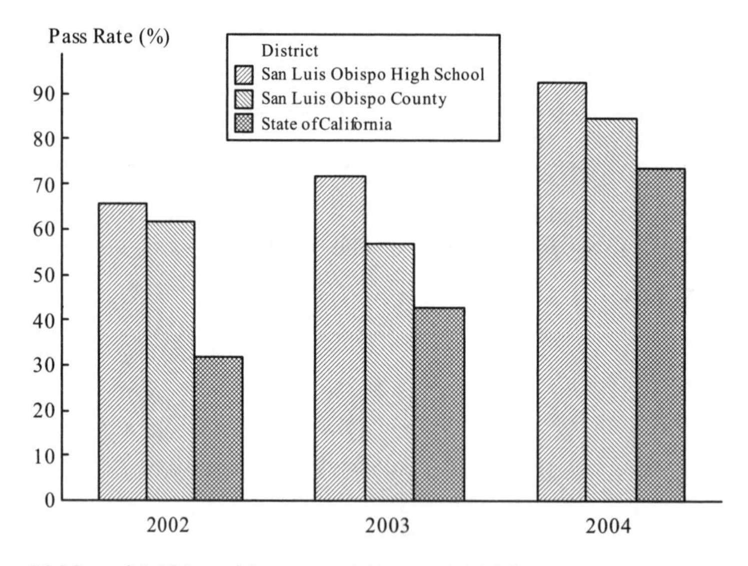

b Between 2002 and 2003 and between 2003 and 2004 the pass rates rose for both the high school and the state, with a particularly sharp rise between 2003 and 2004 for the state. However, for the county the pass rate fell between 2002 and 2003 and then rose between 2003 and 2004.

CR3.11

a

0	123334555599
1	00122234688
2	1112344477
3	0113338
4	37
5	23778

Stem: Thousands
Leaf: Hundreds

The stem-and-leaf display shows a positively skewed distribution with a single peak. There are no extreme values. A typical total length is around 2100 and the great majority of total lengths lie in the 100 to 3800 range.

b

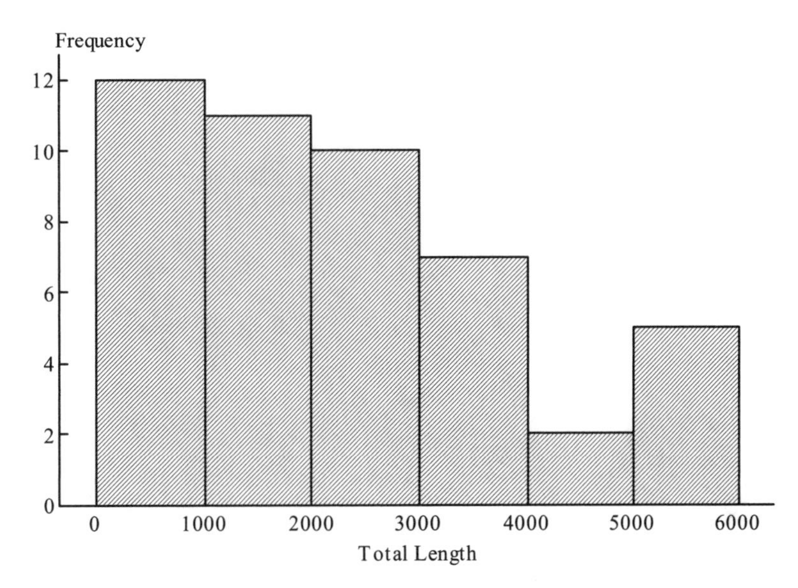

c The number of subdivisions that have total lengths less than 2000 is 12 + 11 = 23, and so the proportion of subdivisions that have total lengths less than 2000 is 23/47 = 0.489.

The number of subdivisions that have total lengths between 2000 and 4000 is 10 + 7 = 17, and so the proportion of subdivisions that have total lengths between 2000 and 4000 is 17/47 = 0.361.

CR3.13

The histogram shows a smooth positively skewed distribution with a single peak. A typical time difference between the two phases of the race is 150 seconds, with the majority of time differences lying between 50 and 350 seconds. There are around three values that could be considered extreme, with those values lying in the 650 to 750 range. Estimating the frequencies from the histogram we see that approximately 920 runners were included in the study and that approximately 8 of those runners ran the late distance more quickly than the early distance (indicated by a negative time difference). Therefore the proportion of runners who ran the late distance more quickly than the early distance is approximately 8/920 = 0.009.

CR3.15

There is a strong negative linear relationship between racket resonance frequency and sum of peak-to-peak accelerations. There are two rackets whose data points are separated from the remaining data points. Those two rackets have very high resonance frequencies and their peak-to-peak accelerations are lower than those of all the other rackets.

Chapter 4
Numerical Methods for Describing Data

4.1 $\bar{x} = (1480 + 1071 + 2291 + 1688 + 1124 + 3476 + 3701)/7$ = $**2118.71**.

To calculate the median, we first list the data values in order of size:

1071 1124 1480 1688 2291 3476 3701

The median is the middle value in this list, $**1688**.

The mean is much larger than the median since the distribution of these seven values is positively skewed. The two largest values are greatly separated from the remaining five values. The median is better as a description of a typical value since it is not influenced by the two extreme values.

4.3 The mean caffeine concentration for the brands of coffee listed is

$$\frac{140+195+155+115+195+180+110+110+130+55+60+60}{12} = \mathbf{125.417} \text{ mg/cup.}$$

Therefore the mean caffeine concentration of the coffee brands in mg/oz is (125.417)/8 = 15.677. This is significantly greater than the mean caffeine concentration of the energy drinks given in the previous exercise.

4.5 The fact that the mean is so much greater than the median tells us that a small number of individuals who donate a large amount of time are greatly increasing the mean.

4.7 **a** There are some unusually large circulation values that make the mean greater than the median.

b The sum of the circulation values given is 13666304, and so the mean is (13666304)/20 = **683315.2**.

The values are already given in descending order, and so to find the median we only need to find the average of the two middle values:
$(438722 + 427771)/2 =$ **433246.5**.

c The median is does the better job of describing a typical value as it is not affected by the small number of unusually large values in the data set.

d This sample is in no way representative of the population of daily newspapers in the US since it consists of the top 20 newspapers in the country.

4.9 **a** The sum of the values given is 8966, and so the mean is 8966/20 = **448.3**.

b Median = (446 + 446)/2 = **446**.

c This sample represents the 20 days with the highest number of speeding-related fatalities, and so it is not reasonable to generalize from this sample to the other 365 days of the year.

4.11 Neither statement is correct. Regarding the first statement it should be noted that, unless the "fairly expensive houses" constitute a majority of the houses selling, these more costly houses will not have an effect on the median. Turning to the second statement, we point out that the small number of very high or very low prices will have no effect on the median, whatever the number of sales. Both statements can be corrected by replacing the median with the mean.

4.13 The two possible solutions are $x_5 = \mathbf{32}$ and $x_5 = \mathbf{39.5}$.

4.15 The two measures of center that can be calculated are the median and the trimmed mean.

To find the median we first list the data values in order:

170 290 350 480 570 790 860 920 1000+ 1000+

The median is the mean of the two middle values: (570 + 790)/2 = **680** hours.

The 20% trimmed mean is (350 + 480 + 570 + 790 + 860 + 920)/6 = **661.667** hours.

4.17 a $\bar{x} = (29+62+37+41+70+82+47+52+49)/9 = 52.111$.

$$\text{Variance} = \frac{(29-52.111)^2 + \cdots + (49-52.111)^2}{8} = \mathbf{279.111}.$$

$s = \sqrt{279.111} = \mathbf{16.707}$.

b The addition of the very expensive cheese would increase the values of both the mean and the standard deviation.

4.19 a The complete data set, listed in order, is:

19 28 30 41 43 46 48 49 53 53 54
62 67 71 77

Lower quartile = 4th value = **41**. Upper quartile = 12th value = **62**. Iqr = **21**.

b The iqr for cereals rated good (calculated in exercise 4.18) is 24. This is greater than the value calculated in Part (a).

4.21 a $\bar{x} = (1480 + \cdots + 3701)/7 = 2118.71429$

Variance $= ((1480 - 2118.71429) + \cdots + (3701 - 2118.71429)^2)/6 = \mathbf{1176027.905}$.

Standard deviation $= \sqrt{1176027.905} = \mathbf{1084.448}$.

The fairly large value of the standard deviation tells us that there is considerable variation between the repair costs.

b For minivans, mean = 1355.833, variance = 93698.967, and standard deviation = 306.103. The mean repair cost for minivans is less than for the smaller cars, showing a lower typical repair cost for the minivans. The standard deviation for minivans is considerably less than for the smaller cars, showing a lower repair cost variability for the minivans.

4.23 **a** The data values, listed in order, are:

0	0	0	0	0	0	0	59	71	83	106
130	142	142	165	177	189	189	189	201	212	224
236	236	306								

Lower quartile = average of 6th and 7th values = (0 + 0)/2 = **0**.
Upper quartile = average of 19th and 20th values = (189 + 201)/2 = **195**.
Interquartile range = 195 − 0 = **195**.

b The lower quartile is equal to the minimum value for this data set because there are a large number of equal values (zero in this case) at the lower end of the distribution. In most data sets this is not the case and therefore, generally speaking, the lower quartile is not equal to the minimum value.

4.25 This data set would have a large standard deviation because parents differ greatly in the amount of money they spend.

4.27 **a** $\bar{x} = (141 + \cdots + 70)/10 = 147.5$.

Variance $= ((141 - 147.5)^2 + \cdots + (70 - 147.5)^2)/9 = 2505.83333$.

Standard deviation $= \sqrt{2505.83333} = \mathbf{50.058}$.

b The Memorial Day data are a great deal more consistent than the New Year's Day data, and therefore the standard deviation for Memorial Day would be smaller than the standard deviation for New Year's Day.

c The standard deviations are given in the table below.

Holiday	Standard Deviation
New Year's Day	50.058
Memorial Day	18.224
July 4th	47.139
Labor Day	17.725
Thanksgiving	15.312
Christmas	52.370

The standard deviations for Memorial Day, Labor Day, and Thanksgiving are 18.224, 17.725, and 15.312, respectively. The standard deviations for the other three holidays are 50.058, 47.139, and 52.370. The standard deviations for the same day of the week holidays are all smaller than all of the standard deviations for the holidays that can occur on different days. There is less variability for the holidays that always occur on the same day of the week.

4.29 **a** The average price for the combined areas would have to take into account the fact that more houses were sold in Los Osos than in Morrow Bay.

b The results for Paso Robles are likely to have the higher standard deviation since the range for Paso Robles (1,575,000 − 170,000 = 1,405,000) is greater than the range for Grover Beach (720,000 − 242,000 = 478,000).

c Assuming that the distributions of house prices are roughly symmetrical, we would expect the median price for Grover Beach to be around (720,000 + 242,000)/2 = 481,000 and the median price for Paso Robles to be around (1,575,000 + 170,000)/2 = 872,500. We expect Paso Robles to have the higher median price.

4.31 a

	Mean	**Standard Deviation**	**Coef. of Variation**
Sample 1	7.81	0.398	5.102
Sample 2	49.68	1.739	3.500

b The values of the coefficient of variation are given in the table in Part (a). The fact that the coefficient of variation is smaller for Sample 2 than for Sample 1 is not surprising since, relative to the actual amount placed in the containers, it is easier to be accurate when larger amounts are being placed in the containers.

4.33 a Median = average of 25th and 26th values = (57.3 + 58.7)/2 = **58**.
Lower quartile = 13th value = **53.5**.
Upper quartile = 38th value = **64.4**.

b (Lower quartile) − 1.5(iqr) = 53.5 − 1.5(10.9) = 37.15.
Since 28.2 and 35.7 are both less than 37.15, they are both outliers.

c

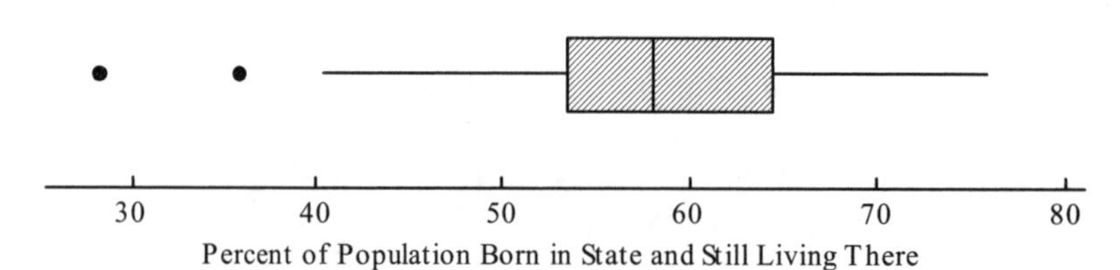

The median percent of population born in the state and still living there is 58. There are two outliers at the lower end of the distribution. If those two values are disregarded the distribution is roughly symmetrical, with values ranging from 40.4 to 75.8.

4.35

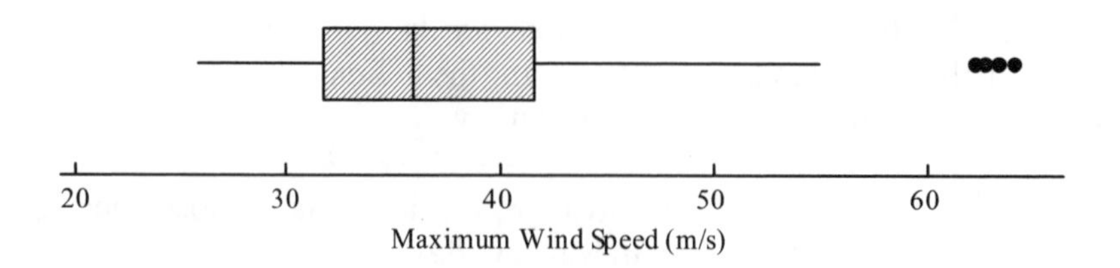

No, the boxplot is not roughly symmetric. It is positively skewed.

4.37 a Since there are outliers in the data set (152 and 43), it would be more appropriate to use the interquartile range than the standard deviation.

b Lower quartile = 81.5, upper quartile = 94, iqr = 12.5.
(Lower quartile) − 3(iqr) = 81.5 − 3(12.5) = 44
(Lower quartile) − 1.5(iqr) = 81.5 − 1.5(12.5) = 62.75
(Upper quartile) + 1.5(iqr) = 94 + 1.5(12.5) = 112.75
(Upper quartile) + 3(iqr) = 94 + 3(12.5) = 131.5

Since the value for students (152) is greater than 131.5, this is an extreme outlier.
Since the value for farmers (43) is less than 44, this is an extreme outlier.
There are no non-extreme outliers.

c

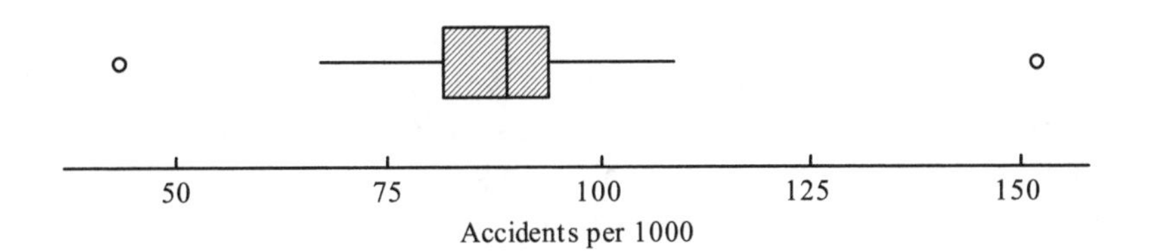

d The insurance company might decide only to offer discounts to occupations that are outliers at the lower end of the distribution, in which case only farmers would receive the discount. If the company was willing to offer discounts to the quarter of occupations with the lowest accident rates then the last 10 occupations on the list should be the ones to receive discounts.

4.39 **a** Since the values given are 1 standard deviation above and below the mean, roughly **68%** of speeds would have been between those two values.

b (1 − 0.68)/2 = 0.16. Roughly **16%** of speeds would exceed 57 mph.

4.41 **a** The values given are two standard deviations below and above the mean. Therefore by Chebyshev's Rule at least 75% of observations must lie between those two values.

b By Chebyshev's Rule at least 89% of observations must lie within 3 standard deviations of the mean. So the required interval is $36.92 \pm 3(11.34) = \mathbf{(2.90, 70.94)}$.

c If the distribution were approximately normal then roughly 2.5% of observations would be more than 2 standard deviations below the mean. However, here $\bar{x} - 2s$ is negative, and so this cannot be the case. Therefore the distribution cannot be approximately normal.

4.43 For the first test $z = (625 - 475)/100 = 1.5$ and for the second test $z = (45 - 30)/8 = 1.875$. Since the student's z score in the second test is higher than in the first, the student did better relative to the other test takers in the second test.

4.45 **a** The values given are 1 standard deviation below and above the mean, so approximately **68%** of the sample observations will be between those values.

b The values given are 2 standard deviations below and above the mean, so approximately **5%** of the sample observations will be outside the interval.

c Approximately (1 − 0.95)/2 = 0.025 of observations lie below 2000 and approximately (1 − 0.68)/2 = 0.16 of observations lie below 2500. Therefore approximately 0.16 − 0.025 = 0.135 (**13.5%**) of observations lie between 2000 and 2500.

d Chebyshev's Rule can only tell us that the required proportions are "at least" something or "at most" something. The Empirical Rule estimates the actual proportions required.

4.47 We require the proportion of observations between 49.75 and 50.25. At 49.75, $z = (49.75 - 49.5)/0.1 = 2.5$. Chebyshev's Rule tells us that at most $1/2.5^2 = 0.16$ of observations lie more than 2.5 standard deviations from the mean. Therefore, since we know nothing about the distribution of weight readings, the best conclusion we can reach is that at most **16%** of weight readings will be between 49.75 and 50.25.

4.49 The value of the standard deviation tells us that a typical deviation of *the number of answers changed from right to wrong* from the mean of this variable is 1.5. However, 0 is only 1.4 below the mean and negative values are not possible, and so for a typical deviation to be 1.5 there must be some values more than 1.5 *above* the mean, that is, values above 2.9. This suggests that the distribution is positively skewed.

The value 6 is the lowest whole number value more than 3 standard deviations above the mean. Therefore, using Chebyshev's Rule, we can conclude that at most $1/3^2 = 1/9$ of students, that is, at most 162/9 = 18 students, changed at least six answers from correct to incorrect.

4.51 **a**

Per Capita Expenditure	Frequency
0 to <2	13
2 to <4	18
4 to <6	10
6 to <8	5
8 to <10	1
10 to <12	2
12 to <14	0
14 to <16	0
16 to <18	2

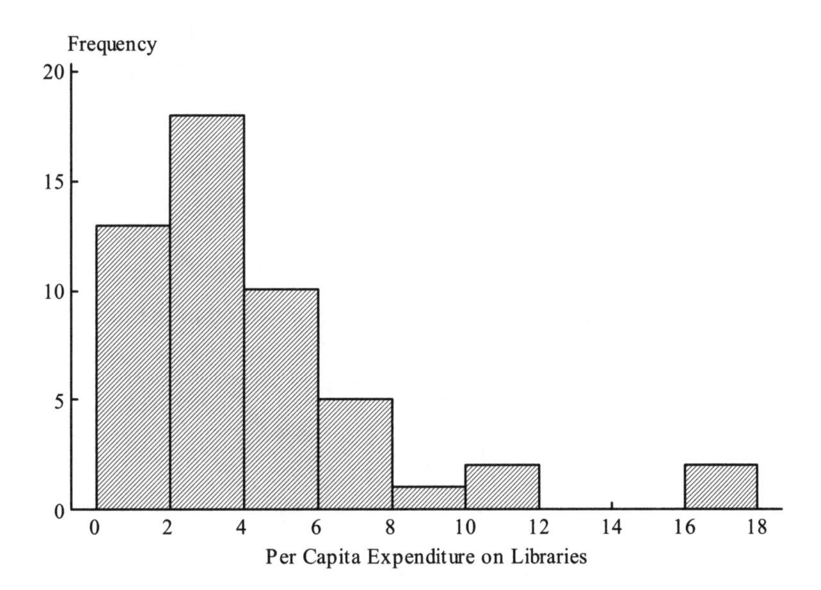

b i **3.4**
ii **5.0**
iii **0.8**
iv **8.0**
v **2.8**

4.53 **a** There is no lower whisker because the minimum value and the lower quartile were both 1.

b The minimum, the lower quartile, and the median are all equal because more than half of the data values were equal to the minimum value.

c The boxplot shows that 2 is between the median and the upper quartile. Therefore between 25% and 50% of patients had unacceptable times to defibrillation.

d (Upper quartile) + 3(iqr) = 3 + 3(2) = 9. Since 7 is less than 9, 7 must be a mild outlier.

4.55 **a** $\bar{x} = (497 + \cdots + 270)/7 = \mathbf{287.714}$.
The seven deviations are
209.286, -94.714, 40.286, -132.714, 38.286, -42.714, -17.714.

b The sum of the rounded deviations is 0.002.

c $\text{Variance} = \left((497 - 287.71429)^2 + \cdots + (270 - 287.71429)^2\right)/6 = \mathbf{12601.905}$.
$s = \sqrt{12601.905} = \mathbf{112.258}$.

4.57 This is the median, and its value is (4443 + 4129)/2 = **$4286**. The other measure of center is the mean, and its value is **$3968.67**. This is smaller than the median and therefore less favorable to the supervisors.

4.59 **a** This is a correct interpretation of the median.

b Here the word "range" is being used to describe the interval from the minimum value to the maximum value. The statement claims that the median is defined to be the midpoint of this interval, which is not true.

c If there is no home below $300,000 then certainly the median will be greater than $300,000 (unless more than half of the homes cost exactly $300,000).

4.61 The new mean is $\bar{x} = (52 + \cdots + 73)/11 = \mathbf{38.364}$.
The new values and their deviations from the mean are shown in the table below.

Value	Deviation
52	13.636
13	-25.364
17	-21.364
46	7.636
42	3.636
24	-14.364
32	-6.364
30	-8.364
58	19.636
35	-3.364

The deviations are the same as the deviations in the original sample. Therefore the value of s^2 for the new values is the same as for the old values. In general, subtracting (or adding) the same number from/to each observation has no effect on s^2 or on s, since the mean is decreased (or increased) by the same amount as the values, and so the deviations from the mean remain the same.

4.63 **a** Lower quartile = 44, upper quartile = 53, iqr = 9.

(Lower quartile) − 1.5(iqr) = 44 − 1.5(9) = 30.5
(Upper quartile) + 1.5(iqr) = 53 + 1.5(9) = 66.5

Since there are no data values less than 30.5 and no data values greater than 66.5, there are no outliers in this data set.

b

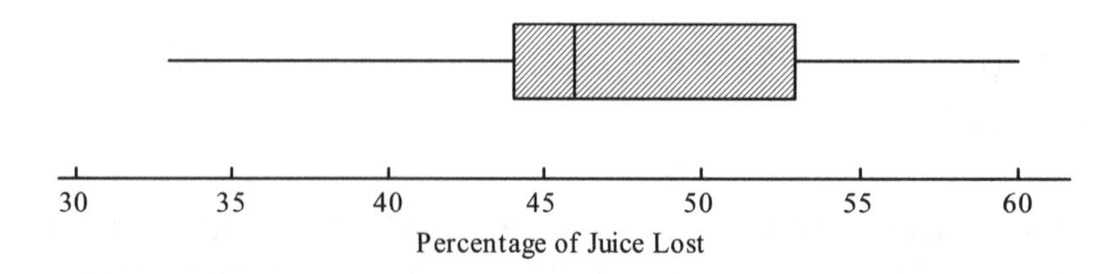

The median of the distribution is 46. The middle 50% of the data range from 44 to 53 and the whole data set ranges from 33 to 60. There are no outliers. The lower half of the middle 50% of data values shows less spread than the upper half of the middle 50% of data values. The spreads of the lowest 25% of data values is slightly greater than the spread of the highest 25% percent of data values.

4.65 **a** $\bar{x} = (244 + \cdots + 200)/14 = \mathbf{192.571}$. This is a measure of center that incorporates all the sample values.

The data values, listed in order, are:

160	174	176	180	180	183	187
191	194	200	205	211	211	244

Median = average of 7th and 8th values = (187 + 191)/2 = **189**. This is a measure of center that is the "middle value" in the sample.

b The mean would decrease and the median would remain the same.

c Trimmed mean $= (174 + \cdots + 211)/12 = \mathbf{191}$.
Trimming percentage = (1/14)(100) = **7.1%**.

d If 244 is changed to 204 then the largest observation is now 211, and one value of 211 will be eliminated from the calculation. This makes the largest three data values in the calculation 204, 205, 211, as compared to 205, 211, 211 in the previous calculation. Therefore the trimmed mean will **decrease**. If 244 is changed to 284, then there is **no change** in the trimmed mean.

4.67

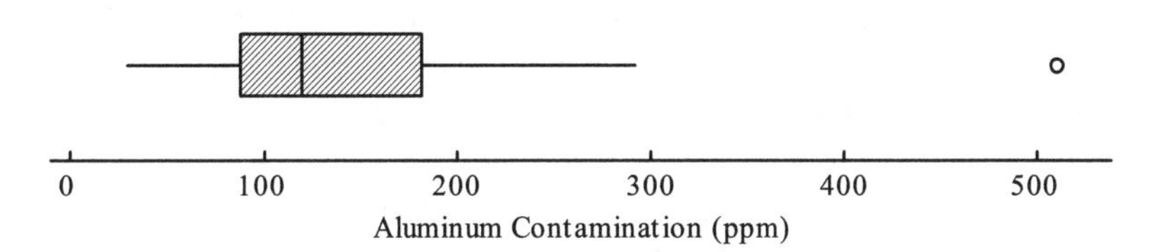

The median aluminum contamination is 119. There is one (extreme) outlier, a value of 511. Disregarding the outlier the data values range from 30 to 291. The middle 50% of data values range from 87 to 182. Even disregarding the outlier the distribution is positively skewed.

4.69

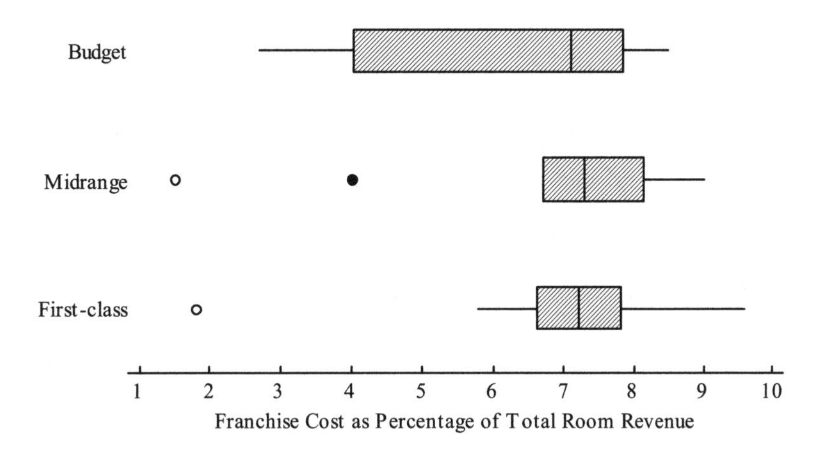

The medians for the three different types of hotel are roughly the same, the median for the midrange hotels being slightly higher than the other two medians. The midrange hotels have two outliers (one extreme) at the lower end of the distribution and the first-class hotels have one (extreme) outlier at the lower end. There are no outliers for the budget hotels. If the outliers are taken into account then the midrange and first-class groups have a greater range than the budget group. If the outliers are disregarded then the budget group has a much greater spread than the other two groups. If the outliers are taken into account then all three distributions are negatively skewed. If the outliers are disregarded then the distribution for the budget group is negatively skewed while the distributions for the other two groups are positively skewed.

4.71 The fact that the mean is greater than the median suggests that the distribution is positively skewed.

4.73 **a** The distribution is roughly symmetrical and 0.84 = 1 − 0.16, and so the 84th percentile is the same distance above the mean as the 16th percentile is below the mean. The 16th percentile is 20 units below the mean and so the 84th percentile is 20 units above the mean. Therefore the 84th percentile is **120**.

b The proportion of scores below 80 is 16% and the proportion above 120 is 16%. Therefore the proportion between 80 and 120 is 100 − 2(16) = 68%. So by the Empirical Rule 80 and 120 are both 1 standard deviation from the mean, which is 100. This tells us that the standard deviation is approximately **20**.

c $z = (90 - 100)/20 = \mathbf{-0.5}$.

d A score of 140 is 2 standard deviations above the mean. By the Empirical Rule approximately 5% of scores are more than 2 standard deviations from the mean. So approximately 5/2 = 2.5% of scores are greater than 140. Thus 140 is at approximately the **97.5th** percentile.

e A score of 40 is 3 standard deviations below the mean, and so the proportion of scores below 40 would be approximately (100 − 99.7)/2 = 0.15%. Therefore there would be very few scores below 40.

Chapter 5
Summarizing Bivariate Data

5.1 **a** Positive. As temperatures increase, cooling costs are likely to increase.

b Negative. As interest rates rise, fewer people are likely to apply for loans.

c Positive. Husbands and wives tend to come from similar backgrounds, and therefore have similar expectations in terms of income.

d Close to zero. There is no reason to believe that there is an association between height and IQ.

e Positive. People with large feet tend to be taller than people with small feet.

f Positive. People who are smart and/or well educated tend to do well on both sections, with those lacking these attributes doing less well on both sections.

g Negative. Those who spend a lot of time on their homework are likely to spend little time watching television, and vice versa.

h Close to zero. The points in the scatterplot will form an inverted "U" shape, making a correlation close to zero.

5.3 Scatterplot for which $r = 1$:

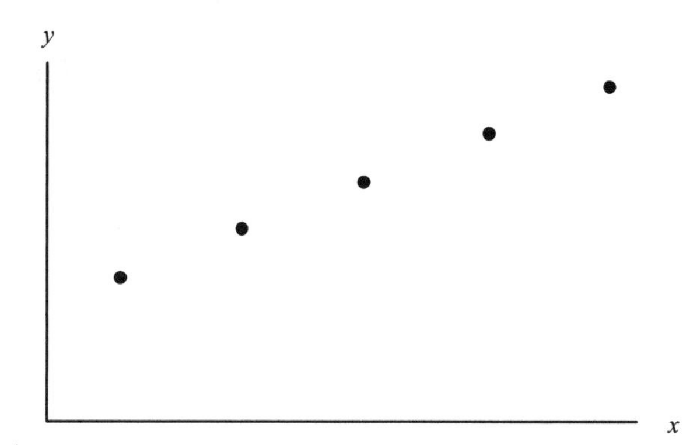

Scatterplot for which $r = -1$:

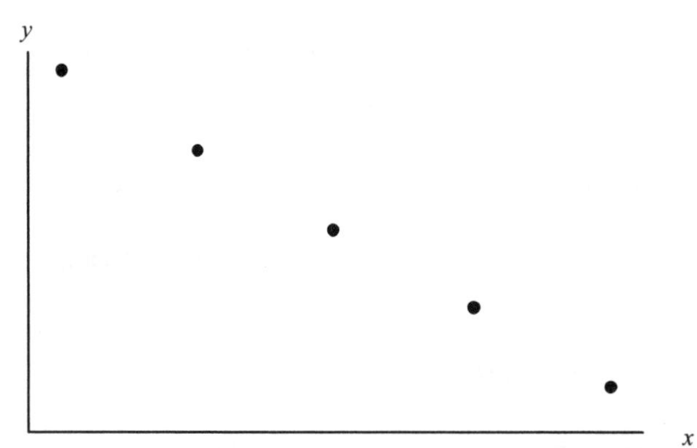

5.5 **a** Using a calculator or statistical software package we get $r =$ **0.204**. There is a weak positive linear relationship between cost per serving and fiber per serving.

b Using a calculator or statistical software package we get $r =$ **0.241**. This correlation coefficient is slightly greater than the correlation coefficient for the per serving data.

5.7 The fact that the correlation coefficient for college GPA and academic self worth was 0.48 tells us that among these athletes there was a weak to moderate positive linear relationship between GPA and self worth. Those with higher grades tended to feel better about themselves in an academic sense than those with lower grades. The correlation coefficient of 0.46 between college GPA and high school GPA gives us the same information about those variables. However, the correlation coefficient of −0.36 between college GPA and the procrastination measure tells us that there was a weak negative linear relationship between those variables. Those who had a tendency to procrastinate generally speaking had lower grades than those without that tendency.

5.9 **a** Using a calculator or statistical software package we get $r =$ **0.118**.

b

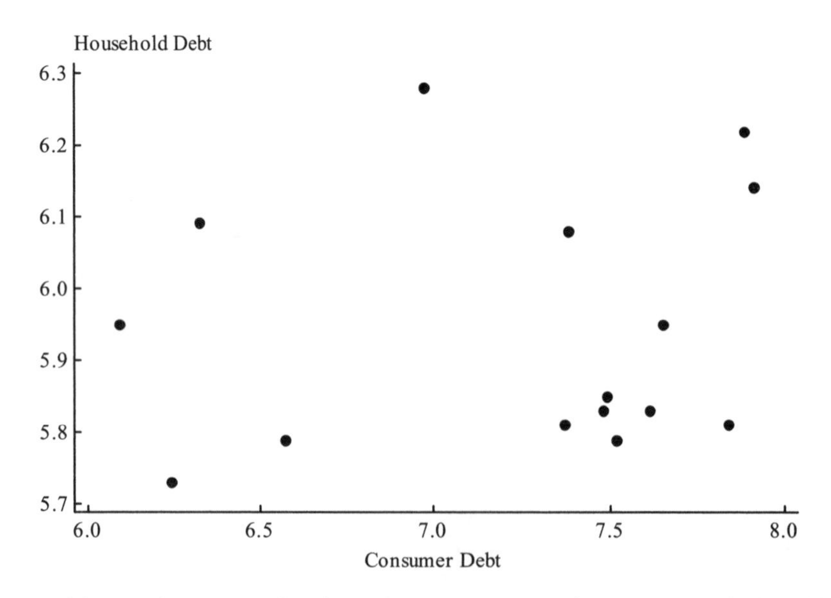

Yes. Looking at the scatterplot there does not seem to be a strong relationship between the variables.

5.11

$$r = \frac{281.1 - \frac{(88.8)(86.1)}{39}}{\sqrt{288 - \frac{88.8^2}{39}}\sqrt{286.6 - \frac{86.1^2}{39}}} = \frac{85.057}{(9.263)(9.824)} = \mathbf{0.935}.$$

There is a strong positive linear relationship between the concentrations of neurolipofuscin in the right and left eye storks.

5.13 The time needed is related to the speed by the equation

$$\text{time} = \frac{\text{distance}}{\text{speed}},$$

where the distance is constant. Using this relationship, and plotting the times (over a fixed distance) for various feasible speeds, a scatterplot is obtained like the one below.

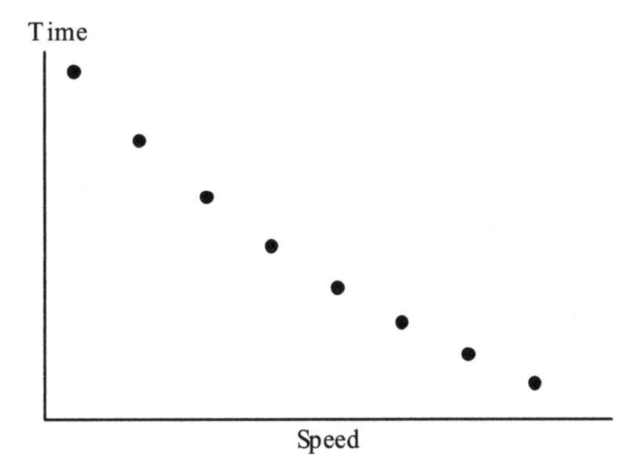

These points show a strong negative correlation, and therefore the correlation coefficient is most likely to be close to −0.9.

5.15 **a**

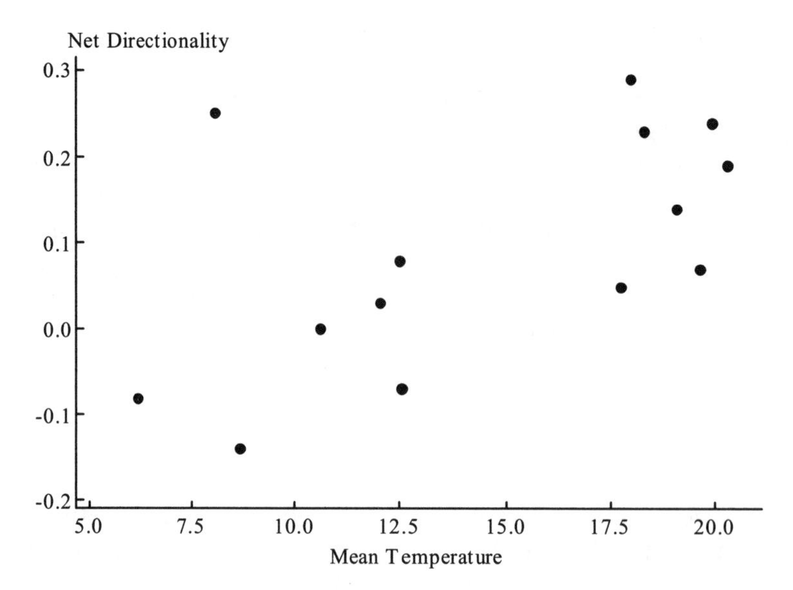

There is one point, (8.06, 0.25), which is separated from the general pattern of the data. If this point is disregarded then there is a somewhat strong positive linear relationship between mean temperature and net directionality. Even if this point is included, there is still a moderate linear relationship between the two variables.

b Using a calculator or statistical software package we find that the equation of the least-squares regression line is $\hat{y} = -0.14282 + 0.016141x$, where x = mean water temperature and y = net directionality.

c When $x = 15$, $\hat{y} = -0.14282 + 0.016141(15) = \mathbf{0.0993}$.

d The scatterplot and the least-squares line support the fact that, generally speaking, the higher the temperature the greater the proportion of larvae that were captured moving upstream.

e Approximately the same number of larvae moving upstream as downstream is represented by a net directionality of zero. According to the least-squares line this will happen when the mean temperature is approximately 8.8°C.

5.17 **a** The dependent variable is the number of fruit and vegetable servings per day, and the predictor variable is the number of hours of television viewed per day.

b Negative. As the number of hours of TV watched per day increases, the number of fruit and vegetable servings per day (on average) decreases.

5.19 **a**

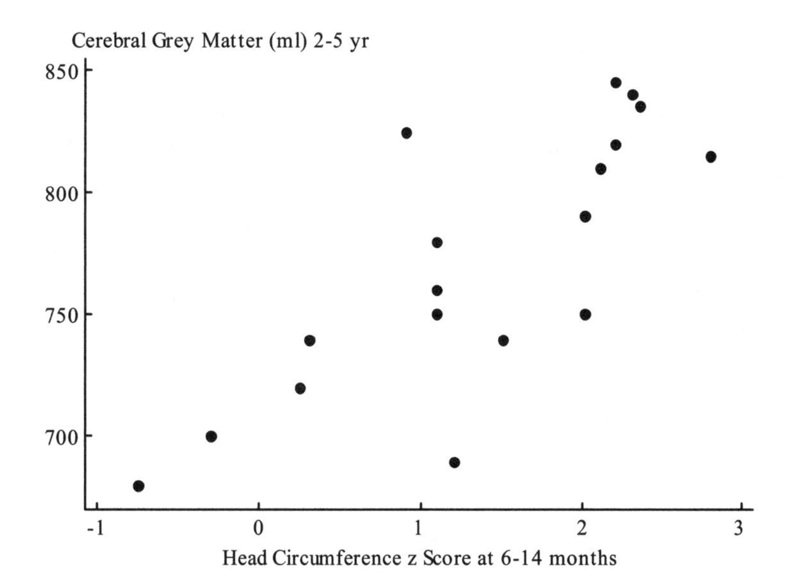

b Using a calculator or statistical software package we find that $r = \mathbf{0.786}$.

c Using a calculator or statistical software package we find that the equation of the least-squares regression line is $\hat{y} = 714.1470 + 42.5196x$, where x = head circumference z score and y = volume of grey matter at 2 to 5 years.

d If $x = 1.8$, then $\hat{y} = 714.1470 + 42.5196(1.8) = \mathbf{790.682}$ ml.

e The value $x = 3.0$ is substantially outside the range of the x-values in the data set, and we do not know that the observed linear pattern continues outside this range. Therefore it would not be a good idea to use the least-squares line to predict the y-value when $x = 3.0$.

5.21 Since the slope of the least-squares line is −9.30, we can say that every extra minute waiting for paramedics to arrive with a defibrillator lowers the chance of survival by 9.3 percentage *points*. (To say that each minute of waiting "lowers the chances of survival by 10 percent" means that one tenth of the probability of surviving is removed for every extra minute of waiting. For example, if the chance of survival after 8 minutes of waiting were 25%, it would mean that the chance of surviving after 9 minutes of waiting was 25 − 2.5 = 22.5%. This is not the case here.)

5.23 **a** Using a calculator or a statistical software package we find that the correlation coefficient between sale price and size is **0.700**. There is a moderate linear relationship between sale price and size.

b Using a calculator or a statistical software package we find that the correlation coefficient between sale price and land-to-building ratio is **−0.332**. There is a weak negative linear relationship between sale price and land-to-building ratio.

c Size is the better predictor of sale price since the absolute value of the correlation between sale price and size is closer to 1 than the absolute value of the correlation between sale price and land-to-building ratio.

d Using a calculator or statistical software package we find that the least-squares regression line for predicting y = sale price from x = size is $\hat{y} = 1.3281 + 0.0053x$.

5.25 The least-squares line is based on the x values contained in the sample. We do not know that the same linear relationship will apply for x values outside this range. Therefore the least-squares line should not be used for x values outside the range of values in the sample.

5.27 We know (as stated in the text) that $b = r(s_y / s_x)$, where s_y and s_x are the standard deviations of the y values and the x values, respectively. Since standard deviations are always positive we know that b and r must always have the same sign.

5.29 **a**

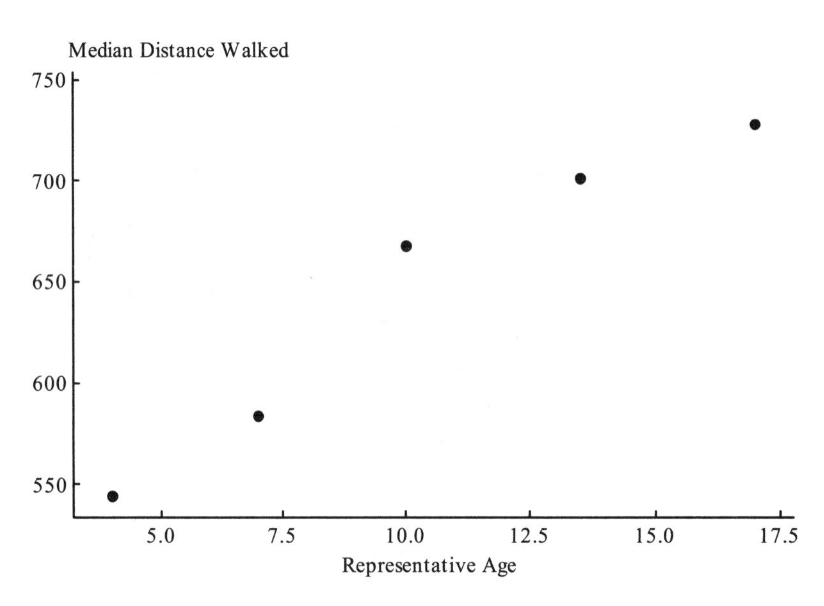

The scatterplot shows a linear pattern between the representative ages of 10 and 17, but there is a greater increase in the median distance walked between the representative ages of 7 and 10 than there is between any other two consecutive age groups.

b Using a calculator or statistical software package we find that the equation of the least-squares regression line is $\hat{y} = 492.79773 + 14.76333x$, where x is the representative age and y is the median distance walked.

c

Representative Age (x)	Median Distance Walked (y)	Predicted y	Residual
4	544.3	551.851	-7.551
7	584	596.141	-12.141
10	667.3	640.431	26.869
13.5	701.1	692.103	8.997
17	727.6	743.774	-16.174

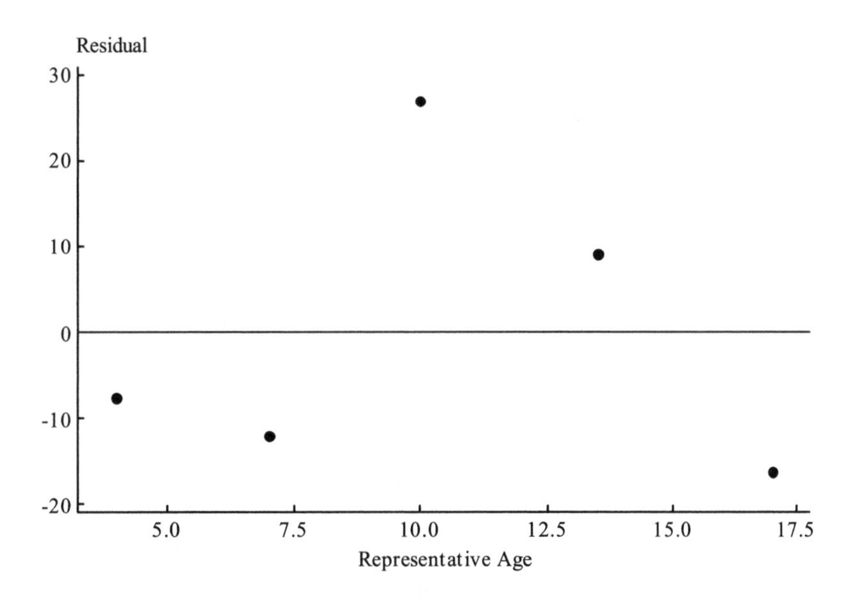

The residual plot reflects the sharp increase in the median distance walked between the representative ages of 7 and 10, with a clear negative residual at $x = 7$ and large positive residual at $x = 10$.

5.31 **a**

Region	**Pollution (x)**	**Medical Cost (y)**	**Predicted y**	**Residual**
North	30	915	941.47	-26.47
Upper South	31.8	891	933.0262	-42.0262
Deep South	32.1	968	931.6189	36.3811
West South	26.8	972	956.4812	15.5188
Big Sky	30.4	952	939.5936	12.4064
West	40	899	894.56	4.44

b $r = \mathbf{-0.581}$. Since the absolute value of r is just a little larger than 0.5 we can describe the linear relationship between pollution and medical cost as moderate.

c

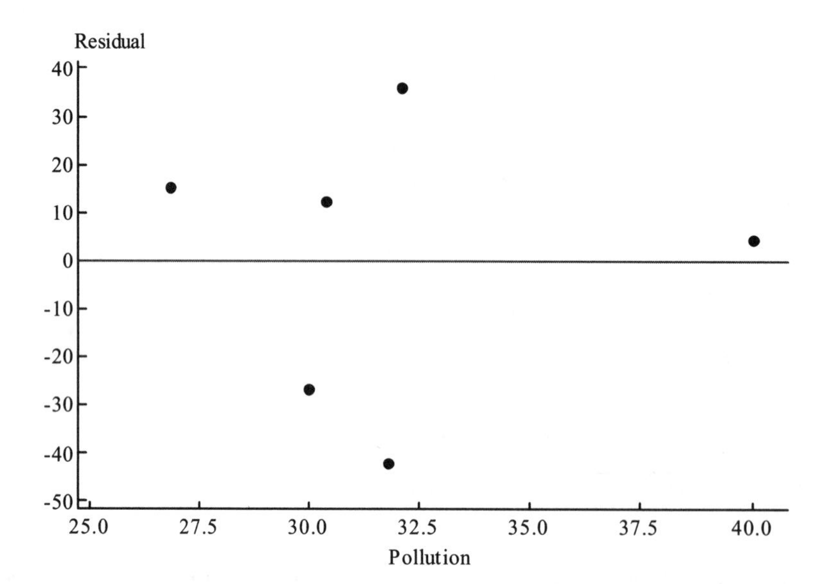

There is one point whose x value is far greater than those of the other points, suggesting that this point might be influential.

d Including the point for the West, the slope of the least-squares line is −4.691 and the intercept is 1082.244. If we remove this point, the resulting slope is −7.107 and the intercept is 1154.371. There is a substantial change in the slope, and therefore the point is influential.

5.33 a

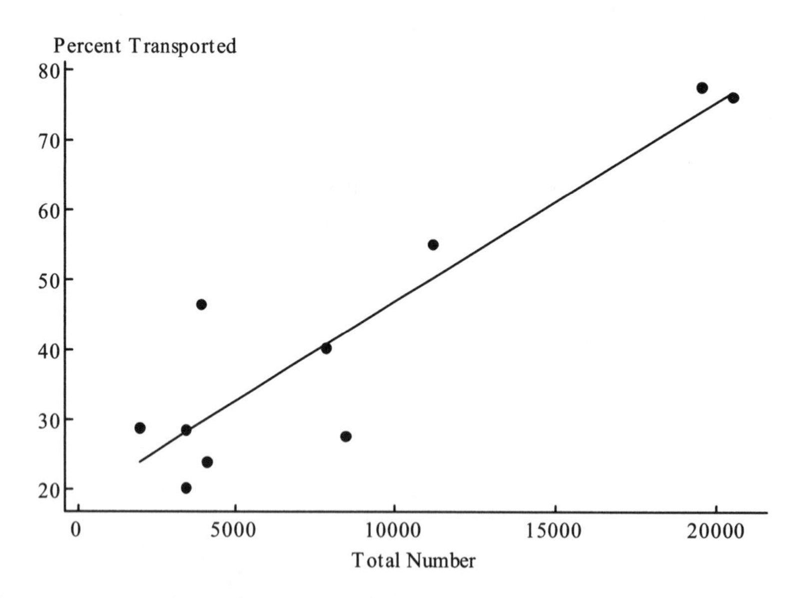

Yes, there appears to be a strong linear relationship between the total number of salmon in the stream and the percent of salmon killed by bears that are transported away from the stream.

b The equation of the least-squares regression line is $\hat{y} = 18.483 + 0.00287x$, where x is the total number of salmon in a creek and y is the percent of salmon killed by bears that were transported away from the stream prior to the bear eating. The regression line has been drawn on the scatterplot in Part (a).

c The point (3928, 46.8) is unlikely to be influential as its x value does not differ greatly from the others in the data set.

d The two points are not influential since the least-squares line provides a good fit for the remaining 8 points. Removing the two points will make only a small change in the regression line.

e $s_e = \mathbf{9.16217}$. This is a typical deviation of a percent transported value from the value predicted by the regression line.

f $r^2 = \mathbf{0.832}$. This is a large value of r^2, and means that 83.2% of the variation in the percent transported values can be attributed to the approximate linear relationship between total number and percent transported.

5.35 Using a calculator or statistical software package we find that $r^2 = \mathbf{0.948}$ and $s_e = \mathbf{20.566}$. The value of r^2 tells us that 94.8% of the variation in six-minute walk time can be attributed to the approximate linear relationship represented by the least-squares line. Since 0.948 is close to 1, the value shows that the fit of the least-squares line to the points is very good. The value of s_e, 20.566, is a typical deviation of a six-minute walk time from the time predicted by the least-squares line.

5.37 **a** The value of r^2 would be **0.154**.

b No, since the r^2 value for y = first year college GPA and x = SAT II score was 0.16, which is not large. Only 16% of the variation in first year college GPA could be attributed to the approximate linear relationship between SAT II score and first year college GPA.

5.39 **a**

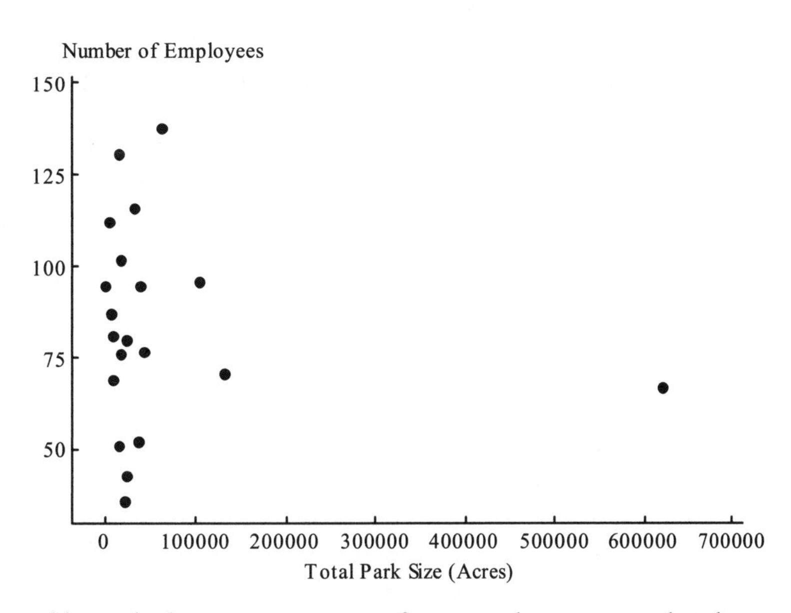

b Using a graphing calculator or computer software package we see that the equation of the least-squares line is $\hat{y} = 85.334 - 0.0000259x$, where x is the total park size in acres and y is the number of employees, and also that the value of r^2 for these two variables is 0.016. With only 1.6% of the variation of the number of employees being attributable to the least-squares line, the line will not give accurate predictions.

c Deleting the point (620231, 67), the equation of the least-squares line is now $\hat{y} = 83.402 + 0.0000387x$. Yes, removal of the point does greatly affect the equation of the line, since it changes the slope from negative to positive.

5.41 The coefficient of determination is $r^2 = 1 - (\text{SSResid/SSTo}) = 1 - (1235.470/25321.368) = \mathbf{0.951}$. This tells us that 95.1% of the variation in hardness is attributable to the approximate linear relationship between time elapsed and hardness.

5.43 **a** The value of r that makes $s_e \approx s_y$ is **0**. The least-squares line is then $\hat{y} = \bar{y}$.

b For values of r close to 1 or −1, s_e will be much smaller than s_y.

c $s_e \approx \sqrt{1-r^2}\,s_y = \sqrt{1-0.8^2}\,(2.5) = \mathbf{1.5}$.

d We now let x = 18-year-old height and y = 6-year-old height. The slope is $b = r(s_y/s_x) = 0.8(1.7/2.5) = 0.544$. So the equation is $\hat{y} = a + bx = a + 0.544x$. The line passes through $(\bar{x}, \bar{y}) = (70, 46)$, so $46 = a + 0.544(70)$, from which we find that $a = 46 - 0.544(70) = 7.92$. Hence the equation of the least-squares line is $\hat{y} = 7.92 + 0.544x$. Also, $s_e \approx \sqrt{1-r^2}\,s_y = \sqrt{1-0.8^2}\,(1.7) = \mathbf{1.02}$.

5.45 **a** The equation of the least-squares quadratic curve is $\hat{y} = 0.8660 - 0.008452x + 0.000410x^2$, where x = percent sunflower meal and y = feed intake.

b When $x = 20$, $\hat{y} = 0.8660 - 0.008452(20) + 0.000410(20)^2 = \mathbf{0.861}$.

5.47 **a**

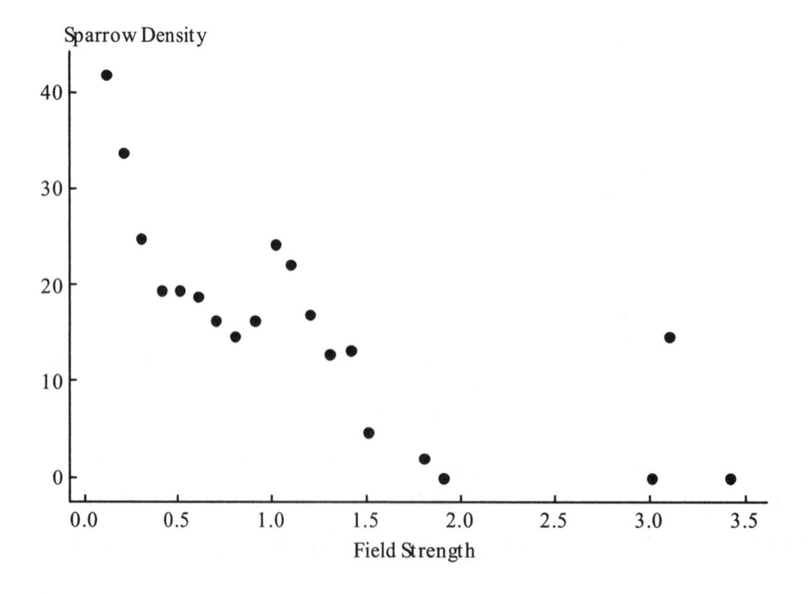

The relationship between sparrow density and field strength appears to be nonlinear.

b When y is plotted against $\sqrt{x}$ the following scatterplot and residual plot are obtained.

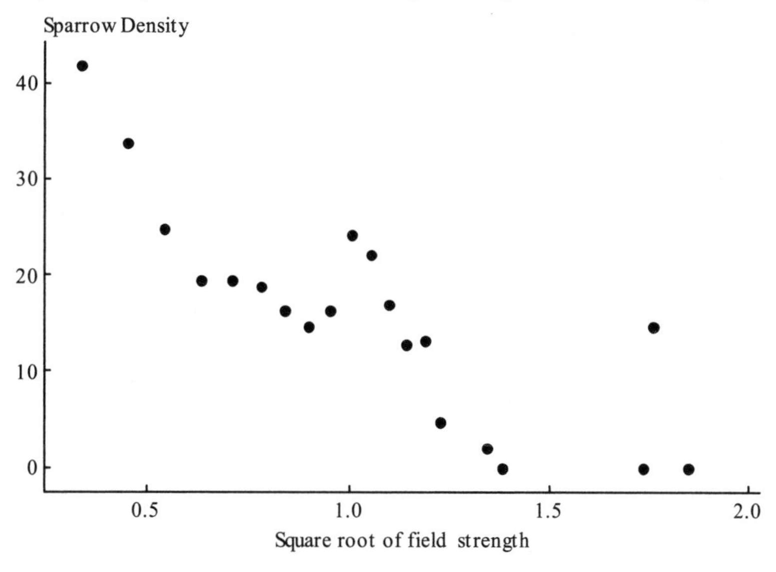

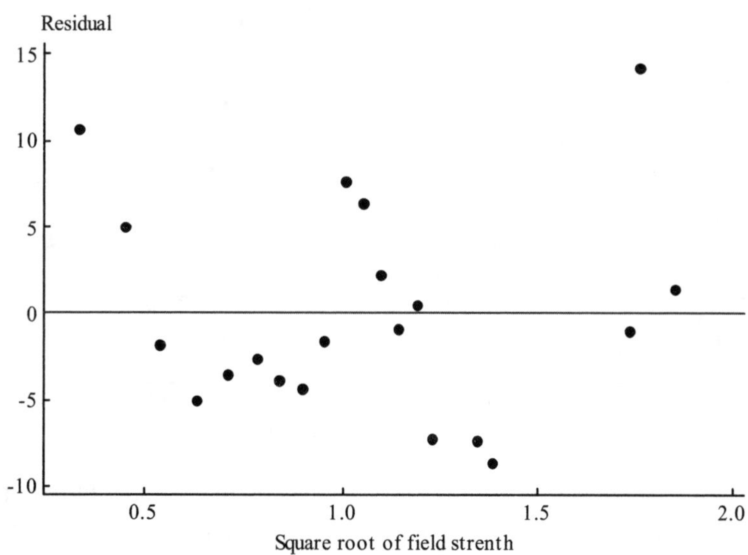

When y is plotted against $\log(x)$ the following scatterplot and residual plot are obtained.

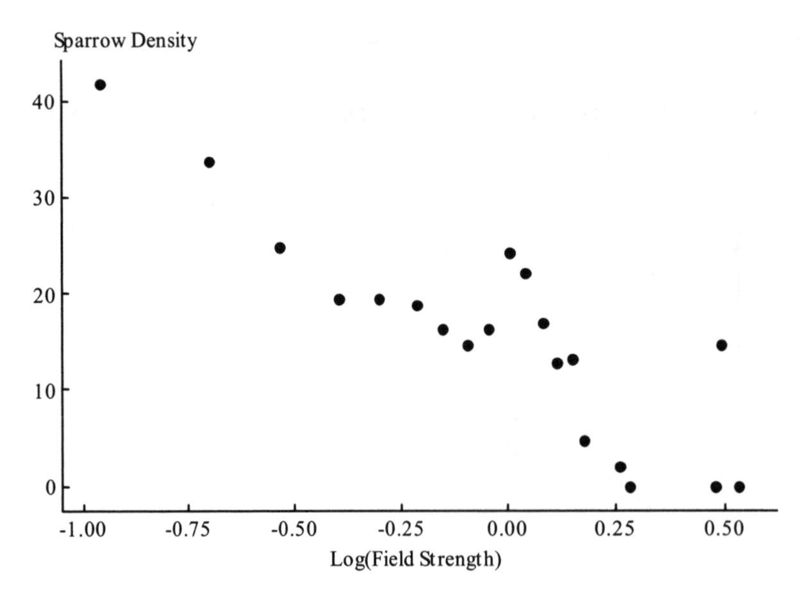

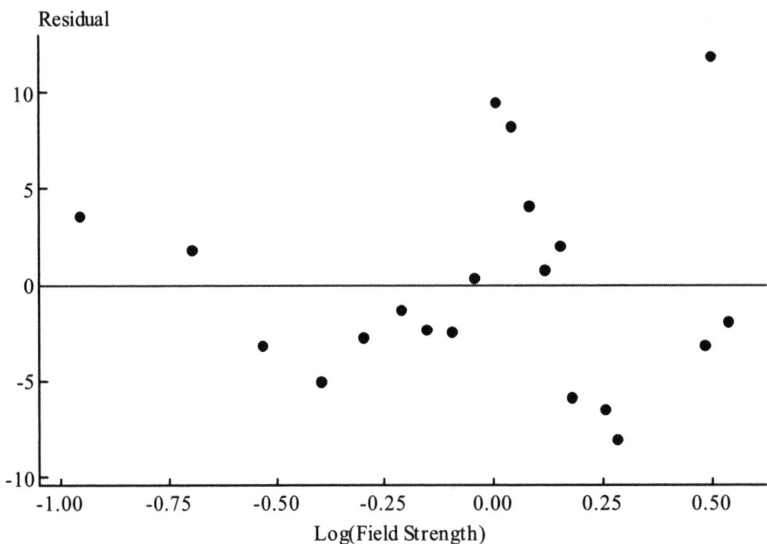

When $x' = \sqrt{x}$ there is slight evidence of a curve in the residual plot, but when $x' = \log(x)$ there is no evidence of a curve in the residual plot. Thus $x' = \log(x)$ is the preferable transformation.

c The equation of the least-squares line is $\hat{y} = 14.80508 - 24.28005 \cdot \log(x)$.

d When $x = 0.5$, $\hat{y} = 14.80508 - 24.28005 \cdot \log(0.5) = \mathbf{22.114}$.
When $x = 2.5$, $\hat{y} = 14.80508 - 24.28005 \cdot \log(2.5) = \mathbf{5.143}$.

5.49 Both x and y have been transformed "down," and a roughly linear pattern has been obtained. Thus a scatterplot of the untransformed data would resemble segment 3 in Figure 5.38.

5.51 **a**

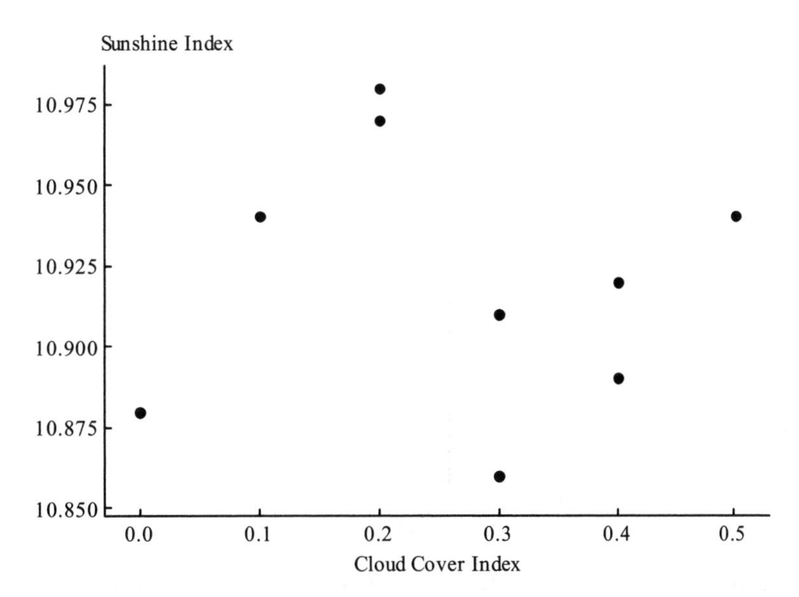

Initially, as the cloud cover index (x) increases from zero, the values of the sunshine index (y) rise. Then, between $x = 0.2$ and $x = 0.3$, the y values seem to decrease sharply, and then to increase again from that point. Certainly neither a linear nor a quadratic model could adequately fit that pattern, however a cubic regression could go some way to modeling the data.

b The least-squares cubic function is $\hat{y} = 10.8768 + 1.4604x - 7.2590x^2 + 9.2342x^3$, where x is the cloud cover index and y is the sunshine index.

c

Cloud Cover Index (x)	Sunshine Index (y)	Residual
.2	10.98	0.02766
.5	10.94	-0.00646
.3	10.91	-0.00088
.1	10.94	-0.01944
.2	10.97	0.01766
.4	10.89	-0.00045
0	10.88	0.00324
.4	10.92	0.02955
.3	10.86	-0.05088

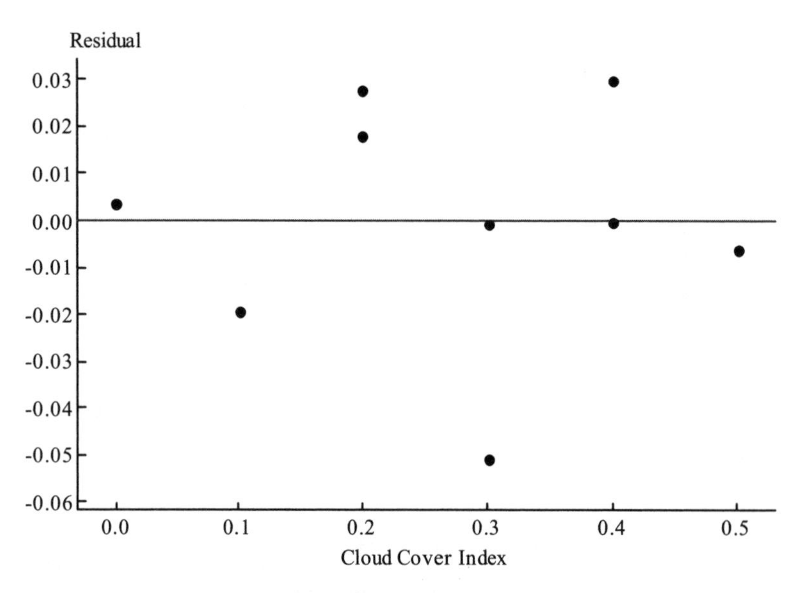

There seems to be a random pattern in the residual plot, suggesting that the cubic regression was appropriate.

d When $x = 0.25$, $\hat{y} = 10.8768 + 1.4604(0.25) - 7.2590(0.25)^2 + 9.2342(0.25)^3 = \mathbf{10.932}$.

e When $x = 0.45$, $\hat{y} = 10.8768 + 1.4604(0.45) - 7.2590(0.45)^2 + 9.2342(0.45)^3 = \mathbf{10.905}$.

f The value 0.75 is well outside the range of the original x values, and we do not know that the cubic relationship that we calculated applies outside this range.

5.53 **a**

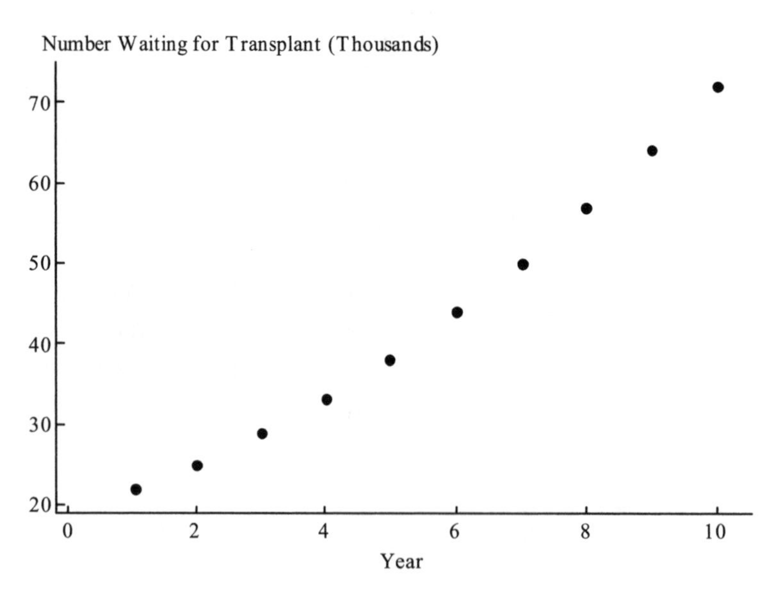

From 1990 to 1999 the number of people waiting for organ transplants increased, with the number increasing by greater amounts each year.

b After much trial and error it is found that the transformation $y' = y^{0.15}$ (with $x' = x$) produces a linear pattern in the scatterplot and a random pattern in the residual plot. The scatterplot is shown below.

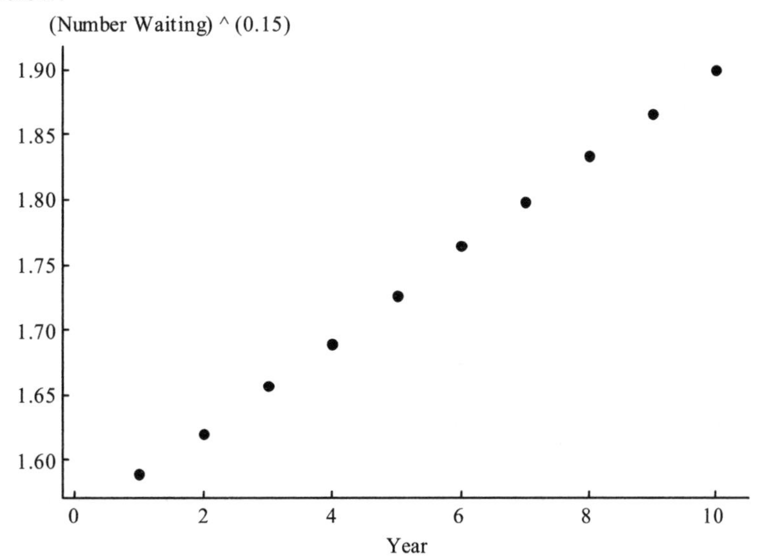

c The least-squares line relating the transformed variables is $\hat{y}^{0.15} = 1.552753 + 0.034856x$, where x is the year (1990 represented by 1) and y is the number waiting (in thousands). When $x = 11$, $\hat{y}^{0.15} = 1.552753 + 0.034856(11) = 1.936164$. From this we get $\hat{y} = (1.936164)^{1/0.15} = \mathbf{81.837}$. The least-squares line predicts that in 2000 the number of patients waiting will be around 81,800.

d We have to be confident that the pattern observed between 1990 and 1999 will continue up to 2000. This is reasonable so long as circumstances remain basically the same. To expect the same pattern to continue to 2010, however, would be unreasonable.

5.55

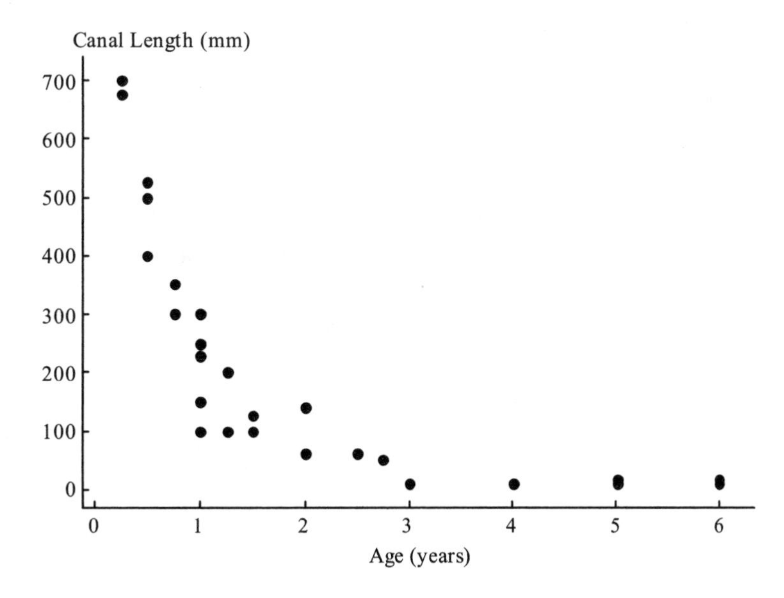

The relationship between age and canal length is not linear. A transformation that makes the plot roughly linear is $x' = 1/\sqrt{x}$ (with $y' = y$). The resulting scatterplot and residual plot are shown below.

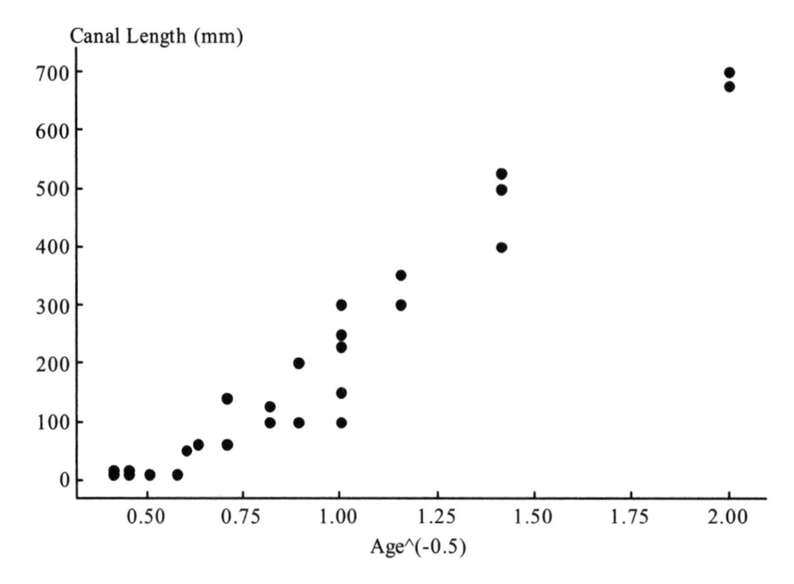

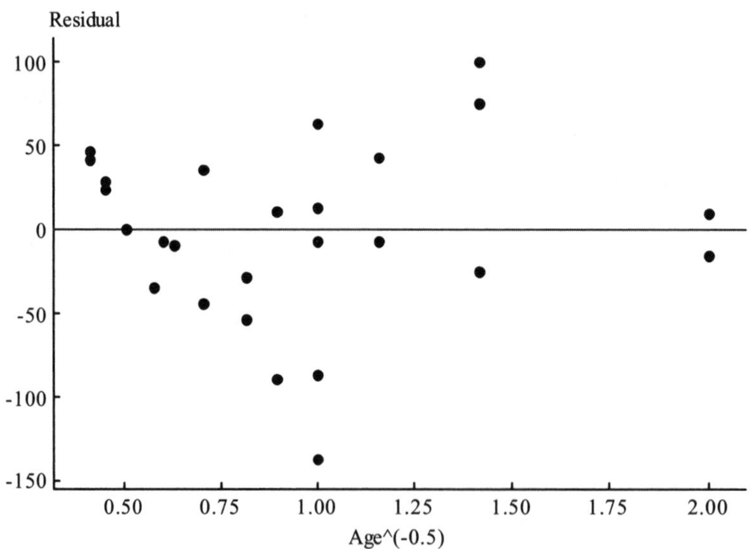

5.57 Calculating the least-squares line for $y' = \ln\left(p/(1-p)\right)$ against x = high school GPA we get $y' = -2.89399 + 1.70586x$. Thus the logistic regression equation is

$$p = \frac{e^{-2.89399+1.70586x}}{1+e^{-2.89399+1.70586x}}.$$

For $x = 2.2$ the equation predicts

$$p = \frac{e^{-2.89399+1.70586(2.2)}}{1+e^{-2.89399+1.70586(2.2)}} = \mathbf{0.702}.$$

5.59 a

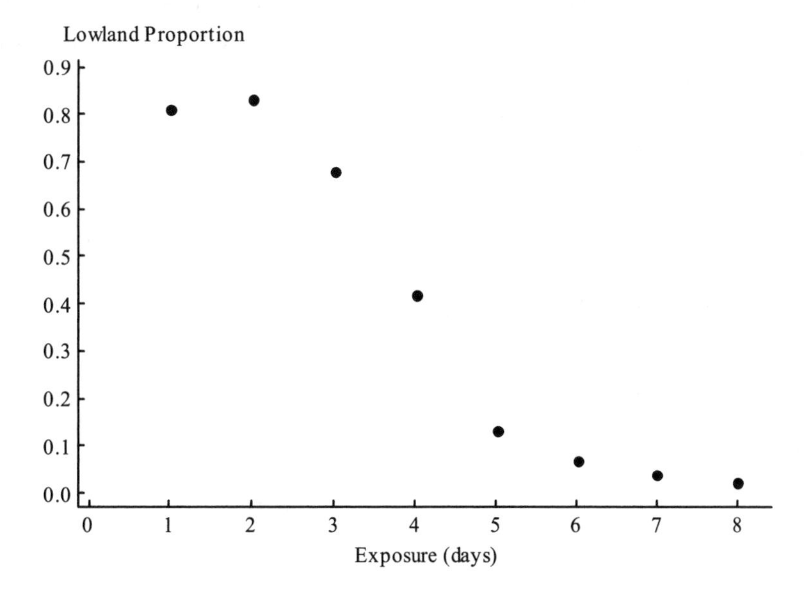

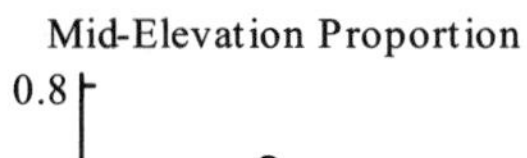

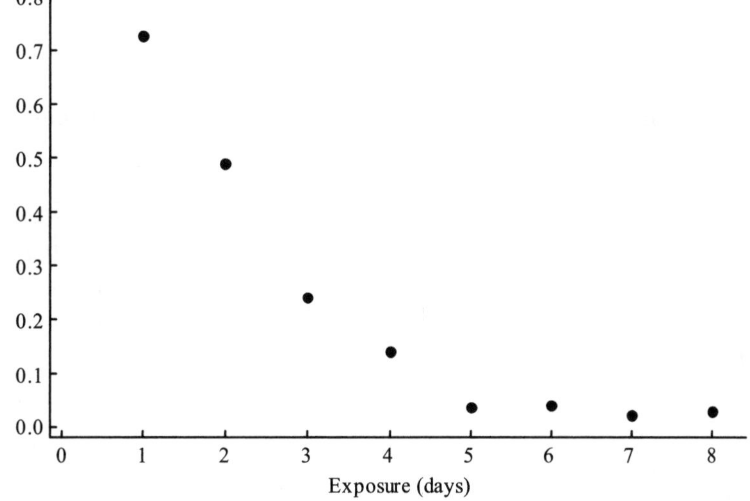

Yes, the plots have roughly the shape you would expect from "logistic" plots.

b

Exposure (days) (x)	**Cloud Forest Proportion (p)**	$y' = \ln(p/(1-p))$
1	0.75	1.09861
2	0.67	0.70819
3	0.36	-0.57536
4	0.31	-0.80012
5	0.14	-1.81529
6	0.09	-2.31363
7	0.06	-2.75154
8	0.07	-2.58669

The least-squares line relating y' and x (where x is the exposure time in days) is $y' = 1.51297 - 0.58721x$. The negative slope reflects the fact that as exposure time increases the hatch rate decreases.

c The logistic regression equation is

$$p = \frac{e^{1.51297-0.58721x}}{1+e^{1.51297-0.58721x}}.$$

For $x = 3$ the equation predicts

$$p = \frac{e^{1.51297-0.58721(3)}}{1+e^{1.51297-0.58721(3)}} = \mathbf{0.438}.$$

For $x = 5$ the equation predicts

$$p = \frac{e^{1.51297-0.58721(5)}}{1+e^{1.51297-0.58721(5)}} = \mathbf{0.194}.$$

d When $p = 0.5$, $y' = \ln(p/(1-p)) = \ln(0.5/(1-0.5)) = 0$. So, solving $1.51297 - 0.58721x = 0$ we get $x = 1.51297/0.58721 = \mathbf{2.577}$ days.

5.61 a

Concentration (g/cc)	Number of Mosquitoes	Number Killed	Proportion Killed	$y' = \ln(p/(1-p))$
0.10	48	10	0.208333	-1.33500
0.15	52	13	0.250000	-1.09861
0.20	56	25	0.446429	-0.21511
0.30	51	31	0.607843	0.43825
0.50	47	39	0.829787	1.58412
0.70	53	51	0.962264	3.23868
0.95	51	49	0.960784	3.19867

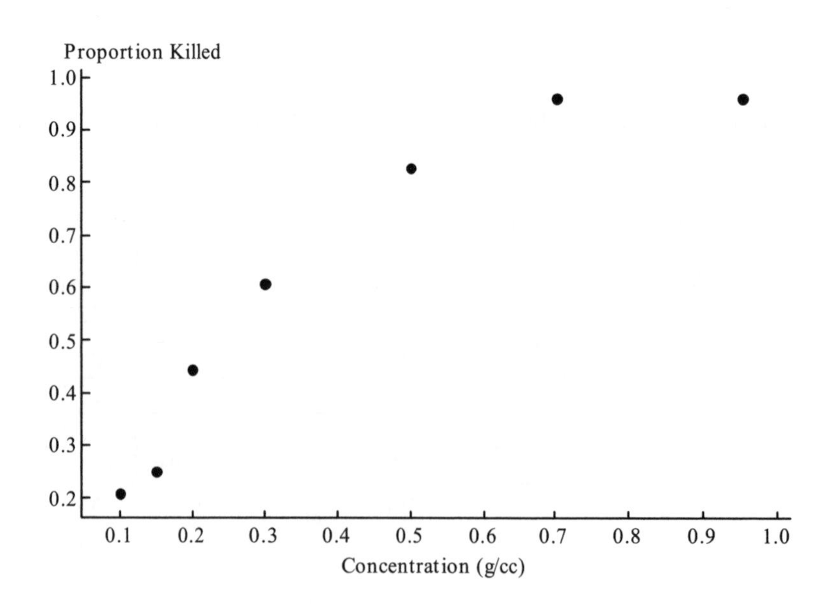

b The least-squares line relating y' and x (where x is the concentration in g/cc) is $\hat{y}' = -1.55892 + 5.76671x$. The positive slope reflects the fact that as the concentration increases the proportion of mosquitoes that die increases.

c When $p = 0.5$, $y' = \ln(p/(1-p)) = \ln(0.5/(1-0.5)) = 0$. So, solving $-1.55892 + 5.76671x = 0$ we get $x = 1.55892/5.76671 = 0.270$. LD50 is estimated to be around **0.270** g/cc.

5.63 **a** Any image plotted between the dashed lines would be associated with Cal Poly by roughly the same percentages of enrolling and non-enrolling students.

b The images that were more commonly associated with non-enrolling students than with enrolling students were "Average," "Isolated," and "Back-up school," with "Back-up school" being the most common of these amongst non-enrolling students. The images that were more commonly associated with enrolling students than with non-enrolling students were (in increasing order of commonality amongst enrolling students) "Excitingly different," "Personal," "Selective," "Prestigious," "Exciting," "Intellectual," "Challenging," "Comfortable," "Fun," "Career-oriented," "Highly respected," and "Friendly," with this last image being marked by over 60% of students who enrolled and over 45% of students who didn't enroll. The most commonly marked image amongst students who didn't enroll was "Career-oriented."

5.65 **a** $r = \sqrt{0.89} = \mathbf{0.943}$. (Note that r is 0.943 rather than −0.943 since the slope of the least-squares line is positive.) There is a very strong positive linear relationship between assault rate and lead exposure 23 years prior. No, we cannot conclude that lead exposure causes increased assault rates, since the value of r close to 1 tells us that there is a strong linear *association* between lead exposure and assault rate, but tells us nothing about *causation*.

b The equation of the least-squares regression line is $\hat{y} = -24.08 + 327.41x$, where y is the assault rate and x is the lead exposure 23 years prior. When $x = 0.5$, $\hat{y} = -24.08 + 327.41(0.5)$ $= \mathbf{139.625}$ assaults per 100,000 people.

c **89%** of the year-to-year variability in assault rates can be explained by the relationship between assault rate and gasoline lead exposure 23 years earlier.

d The two time series plots, generally speaking, move together. That is, generally when one goes up the other goes up and when one goes down the other goes down. Thus high assault rates are associated with high lead exposures 23 years earlier and low assault rates are associated with low lead exposures 23 years earlier.

5.67 **a** $r = \mathbf{-0.981}$. This suggests a very strong linear relationship between the amount of catalyst and the resulting reaction time.

b

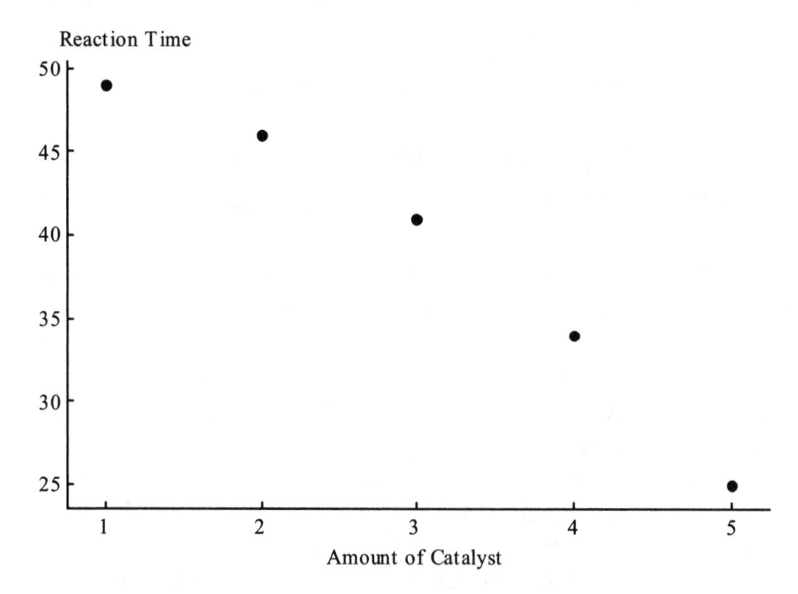

The word *linear* does not provide the most effective description of the relationship. There are curves that would provide a much better fit.

5.69 **a**

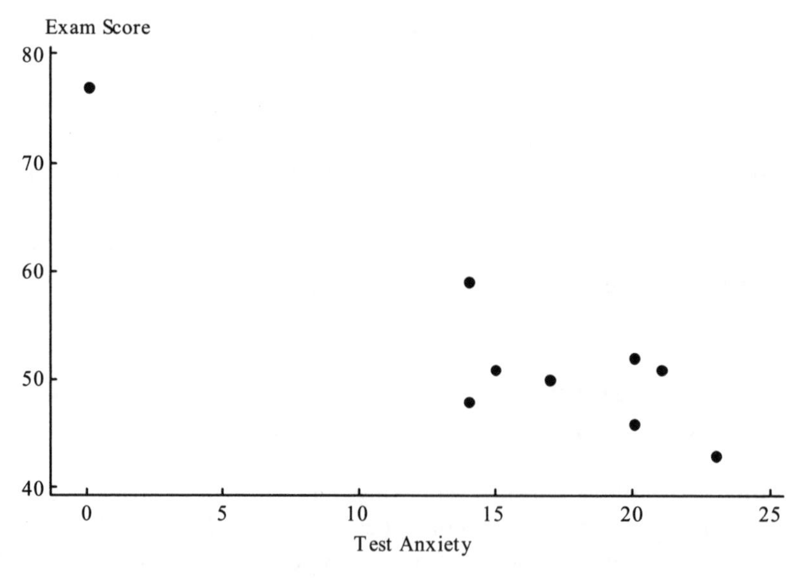

There is one point, (0, 77), that is far separated from the other points in the plot. There is a clear negative relationship between scores on the measure of test anxiety and exam scores.

b There appears to be a very strong negative linear relationship between test anxiety and exam score. (However, without the point (0, 77) the relationship would be significantly less strong.)

c $r = \mathbf{-0.912}$. This is consistent with the observations given in Part (b).

d No, we cannot conclude that test anxiety caused poor exam performance. Correlation measures the strength of the linear relationship between the two variables, but tells us nothing about causation.

5.71 **a**

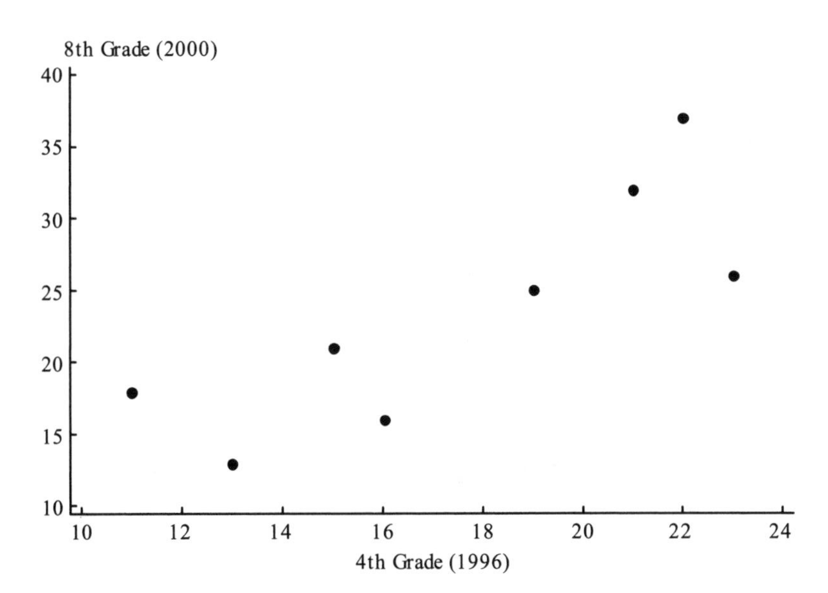

There is a clear positive relationship between the percentages of students who were proficient at the two times. There is the suggestion of a curve in the plot.

b The equation of the least-squares line is $\hat{y} = -3.13603 + 1.52206x$, where x is the percentage proficient in 4th grade (1996) and y is the percentage proficient in 8th grade (2000).

c When $x = 14$, $\hat{y} = -3.13603 + 1.52206(14) = \mathbf{18.173}$. This is slightly lower than the actual value of 20 for Nevada.

5.73 **a** When $x = 25$, $\hat{y} = 62.9476 - 0.54975(25) = 49.204$. So the residual is $y - \hat{y} = 70 - 49.204 = \mathbf{20.796}$.

b $r = -\sqrt{0.57} = \mathbf{-0.755}$ (The correlation coefficient is negative since the slope of the least-squares regression line is negative.)

c We know that $r^2 = 1 - \text{SSResid}/\text{SSTo}$. Solving for SSResid we get $\text{SSResid} = \text{SSTo}(1 - r^2) = 2520(1 - 0.57) = 1083.6$. Therefore $s_e = \sqrt{\text{SSResid}/(n-2)} = \sqrt{1083.6/8} = \mathbf{11.638}$.

5.75 **a** $r = \mathbf{-0.717}$

b $r = \mathbf{-0.835}$. The absolute value of this correlation is greater than the absolute value of the correlation calculated in Part (a). This suggests that the transformation was successful in straightening the plot.

5.77 a

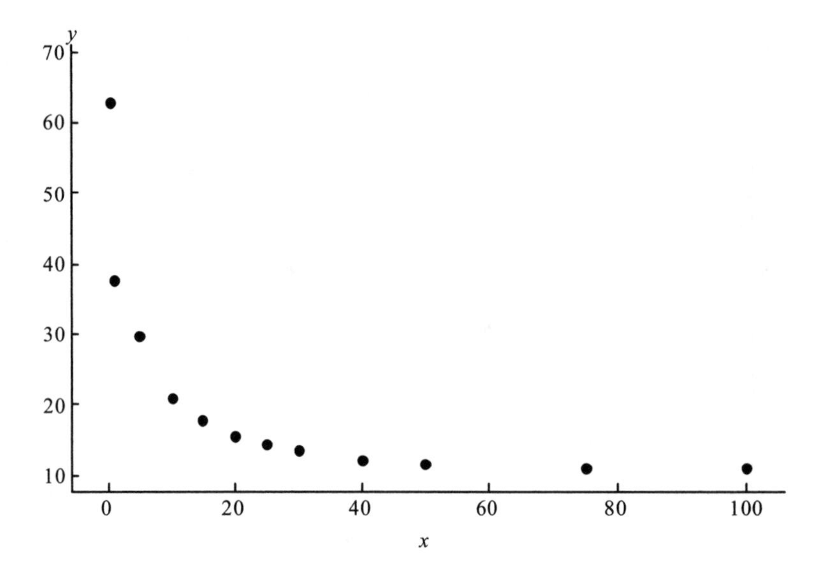
y
70
60
50
40
30
20
10
0
20
40
60
80
100
x

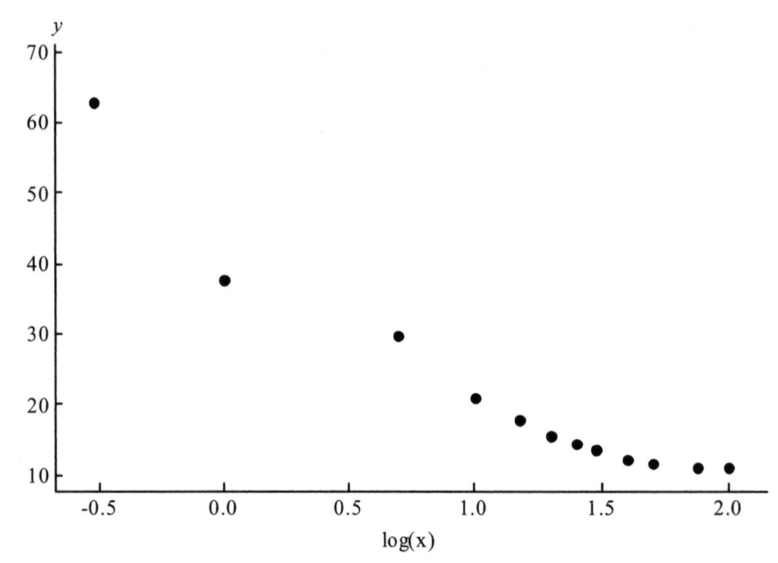
y
70
60
50
40
30
20
10
-0.5
0.0
0.5
1.0
1.5
2.0
log(x)

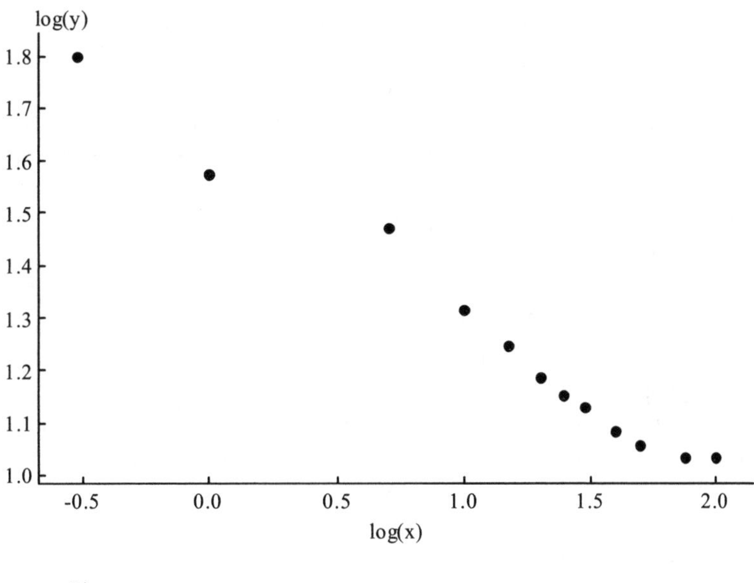

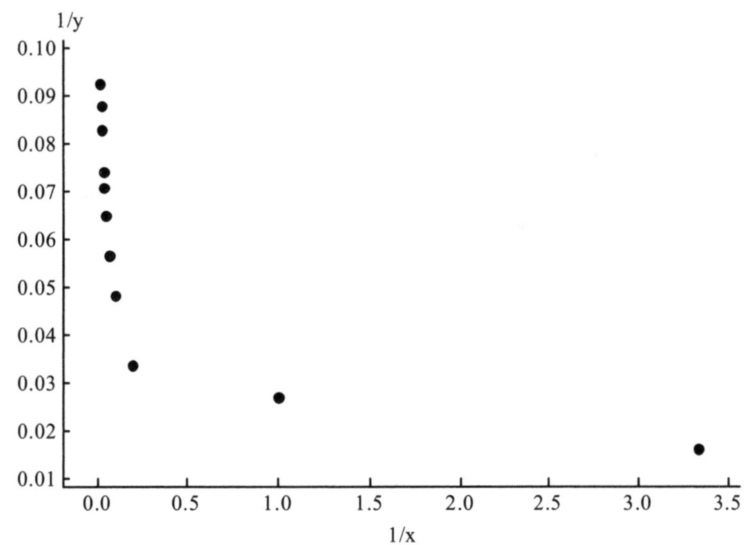

b Plotting $\log(y)$ against $\log(x)$ does the best job of producing an approximately linear relationship. The least-squares line of $\log(y)$ on $\log(x)$ is $\log(\hat{y}) = 1.61867 - 0.31646\log(x)$. So when $x = 25$, $\log(\hat{y}) = 1.61867 - 0.31646\log(25) = 1.17629$. Therefore $\hat{y} = 10^{1.17629} = 15.007$. The predicted lead content is **15.007** parts per million.

5.79 **a** $r = 0$

b For example, adding the point (6, 1) gives $r = 0.510$. (Any y-coordinate greater than 0.973 will work.)

c For example, adding the point (6, −1) gives $r = -0.510$. (Any y-coordinate less than −0.973 will work.)

Cumulative Review Exercises

CR5.1 Here is one possible design. Gather a number of volunteers (around 50, for example) who are willing to take part in an experiment involving exercise. Establish some measure of fitness, involving such criteria as strength, endurance, and muscle mass. Measure the fitness of each person. Randomly assign the 50 people to two groups, Group A and Group B. (This can be done by writing the names of the 50 people on identical slips of paper, placing the slips of paper in a hat, mixing them, and picking 25 names at random. Those 25 people will be put into Group A and the remainder will be put into Group B.) People in Group A should be instructed on a program of exercise that does not involve the sort of activity one would engage in at the gym, and this exercise should be undergone wearing the new sneakers. People in Group B should be instructed on an equivalent program of exercise that primarily involves gym-based activities, and this exercise should be undergone without the wearing of the new sneakers. At the end of the program the fitness of all the participants should be measured and a comparison should be made regarding the increase in fitness of the people in the two groups.

This is an experiment since the participants are assigned to the groups by the experimenters.

CR5.3 The peaks in rainfall do seem to be followed by peaks in the number of E. coli cases, with rainfall peaks around May 12, May 17, and May 23 being followed by peaks in the number of cases on May 17, May 23, and May 28th. (The incubation period seems to be more like 5 days than the 3 to 4 days mentioned in the caption.) Thus the graph does show a close connection between unusually heavy rainfall and the incidence of the infection. The storms may not be *responsible* for the increased illness levels, however, since the graph can only show us association, not causation.

CR5.5

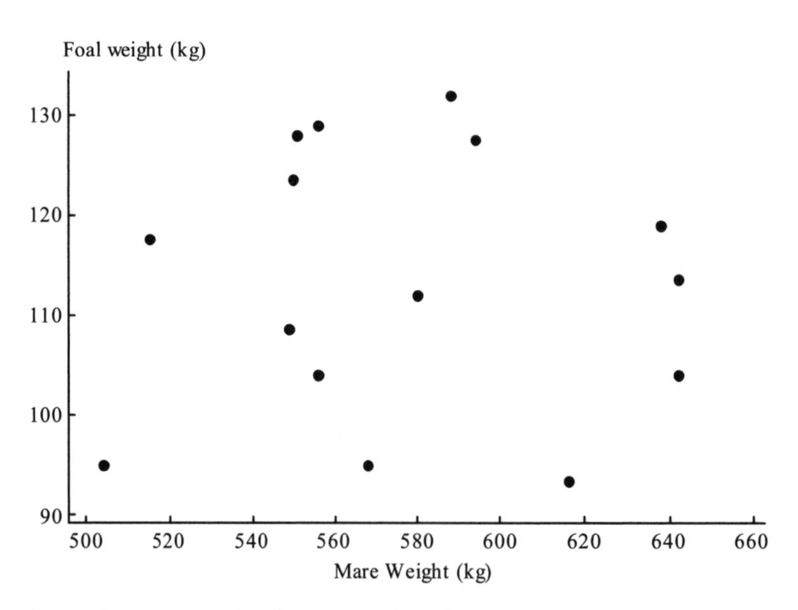

The apparently random pattern in the scatterplot shows that there is very little relationship between the weight of the mare and the weight of her foal. This is supported by the value of the correlation coefficient. A value so close to zero shows that there is little to no linear relationship between the weight of the mare and the weight of the foal.

CR5.7 a

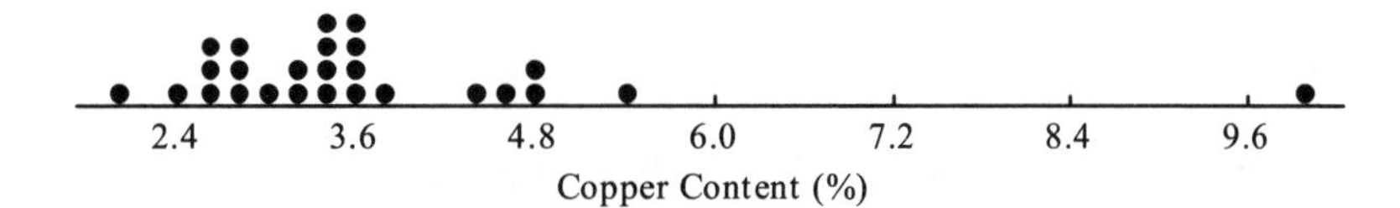

b $\bar{x} = (2.0 + \cdots + 10.1)/26 = 3.654$. The mean copper content is **3.654%**.
Median = average of 13th and 14th values = (3.3 + 3.4)/2 = 3.35. The median copper content is **3.35%**.

c With a sample size of 26, the 8% trimmed mean removes 2 values from each end, since 8% of 26 is approximately 2. Removing 10.1 and 5.3 from the upper end will result in a noticeable reduction in the mean since 10.1 is an extreme value, while removing 2.0 and 2.4 from the lower end will have less effect on the mean. Therefore the trimmed mean will be smaller than the mean.

CR5.9 a The dotplot and stem-and-leaf display are shown below.

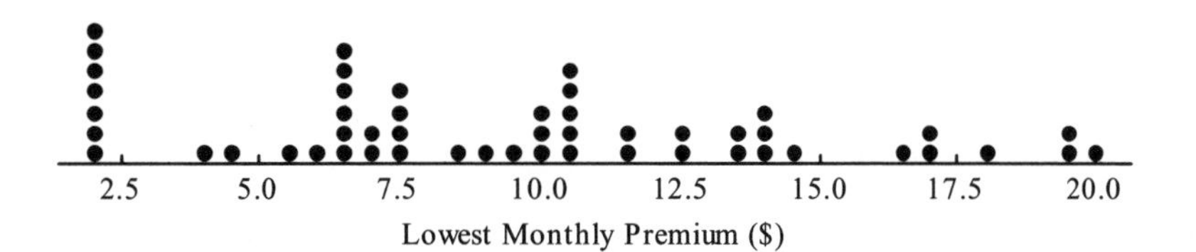

Stem	Leaf
1	8888888
2	
3	
4	14
5	4
6	133444499
7	3333
8	68
9	4
10	01123336
11	46
12	33
13	237
14	004
15	
16	5
17	019
18	
19	66
20	0

Stem: Ones
Leaf: Tenths

b Looking at the displays, one would expect the mean and the median to be roughly the same. (Looking at the data points between 4.1 and 20.0, you might notice some positive skewness, and therefore conclude that the mean would be bigger than the median. However, the seven values of 1.87 separated from the rest of the data at the lower end of the distribution will roughly compensate for that positive skew making the mean and the median roughly equal.)

c Mean = **\$9.459**, median = **\$9.48.**

d A dotplot for the highest premium data is shown below.

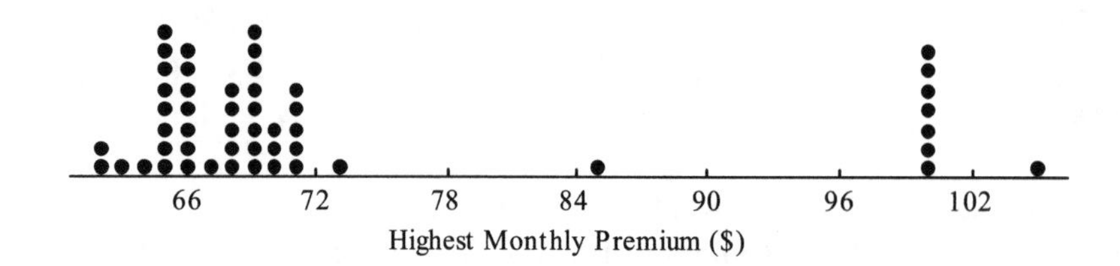

e Mean = **\$72.846**, median = **\$68.61**.

CR5.11

a $\bar{x} = (3099 + \cdots + 3700)/10 = \mathbf{2965.2}$.

$\text{Variance} = \left((3099 - 2965.2)^2 + \cdots + (3700 - 2965.2)^2\right)/9 = 294416.622$.

$s = \sqrt{294416.622} = \mathbf{542.602}$.

The data values listed in order are:

2297 2401 2510 2682 2824 3068 3099 3112 3700 3959

Lower quartile = 3rd value = 2510.
Upper quartile = 8th value = 3112.
Interquartile range = 3112 − 2510 = **602**.

b The interquartile range for the chocolate pudding data (602) is less than the interquartile range for the tomato catsup data (1300). So there is less variability in sodium content for the chocolate pudding data than for the tomato catsup data.

CR5.13

a $\bar{x} = (4.8 + \cdots + 3.7)/20 = \mathbf{4.93}$.

The data, listed in order are:

0.4	0.9	1.4	1.4	2.1	2.4	2.9	3.3	3.4	3.5	3.7
4.8	5	5	5.4	6.1	7.5	10.8	13.8	14.8		

Median = average of 10th and 11th = (3.5 + 3.7)/2 = **3.6**.

The mean is greater than the median. This is explained by the fact that the distribution of blood lead levels is positively skewed.

b

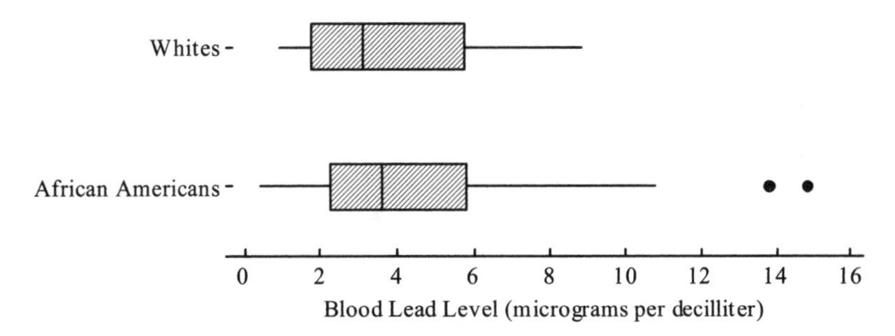

The median blood lead level for the African Americans (3.6) is slightly higher than for the Whites (3.1). Both distributions seem to be positively skewed. There are two outliers in the data set for the African Americans. The distribution for the African Americans shows a greater range than the distribution for the Whites, even if you discount the two outliers.

CR5.15

a Yes, it appears that the variables are highly correlated.

b There is a strong positive linear relationship between the observations by the standard spectrophotometric method and the new, simpler method.

c Perfect correlation would result in the points lying exactly on some straight line, but not necessarily on the line described.

CR5.17

a This value of r^2 tells us that 76.64% of the variability in clutch size can be attributed to the approximate linear relationship between snout-vent length and clutch size.

b Using $r^2 = 1 - \text{SSResid/SSTo}$ we see that $\text{SSResid} = \text{SSTo}(1 - r^2)$. So here $\text{SSResid} = 43951(1 - 0.7664) = 10266.9536$. Therefore

$s_e = \sqrt{\dfrac{\text{SSResid}}{n-2}} = \sqrt{\dfrac{10266.9536}{12}} = \mathbf{29.250}$. This is a typical deviation of an observed clutch size from the clutch size predicted by the least-squares line.

CR5.19

a

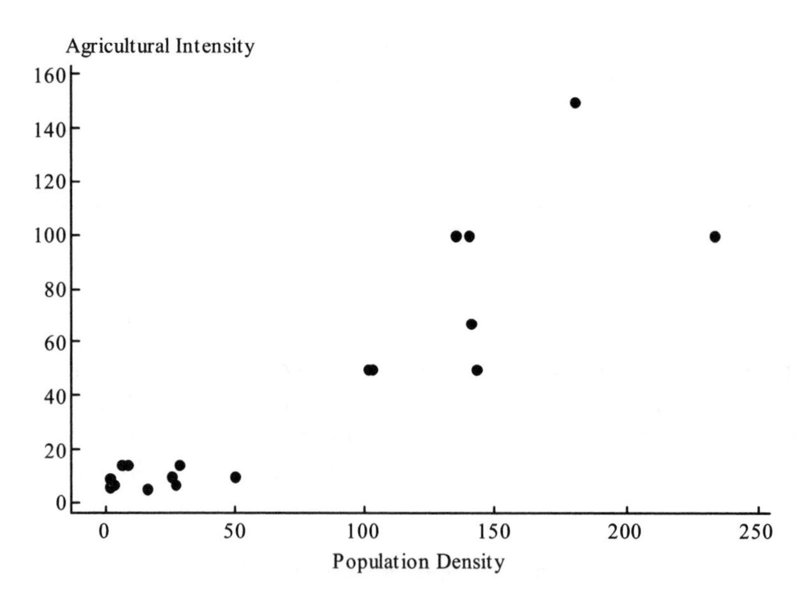

Yes, the scatterplot shows a strong positive association between population density and agricultural intensity.

b

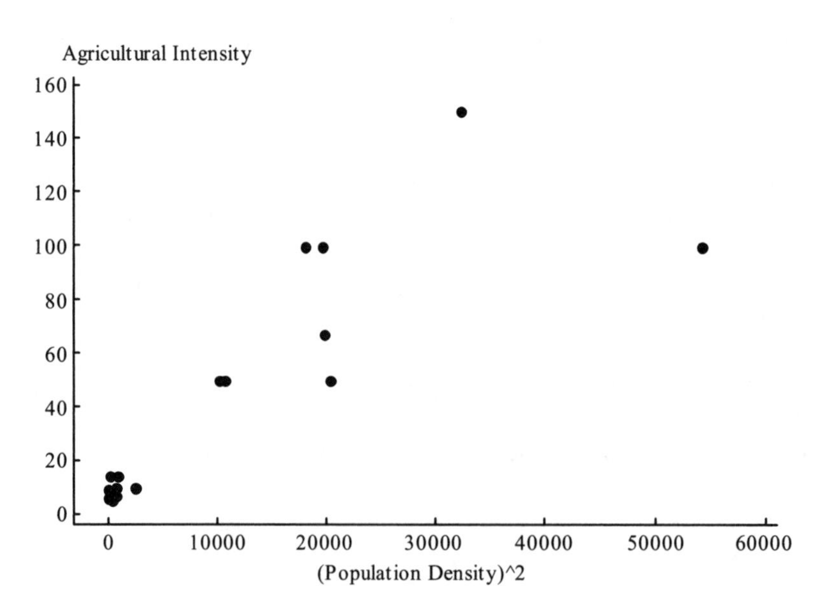

The plot now seems to be straight, particularly if you disregard the point with the greatest x value.

c

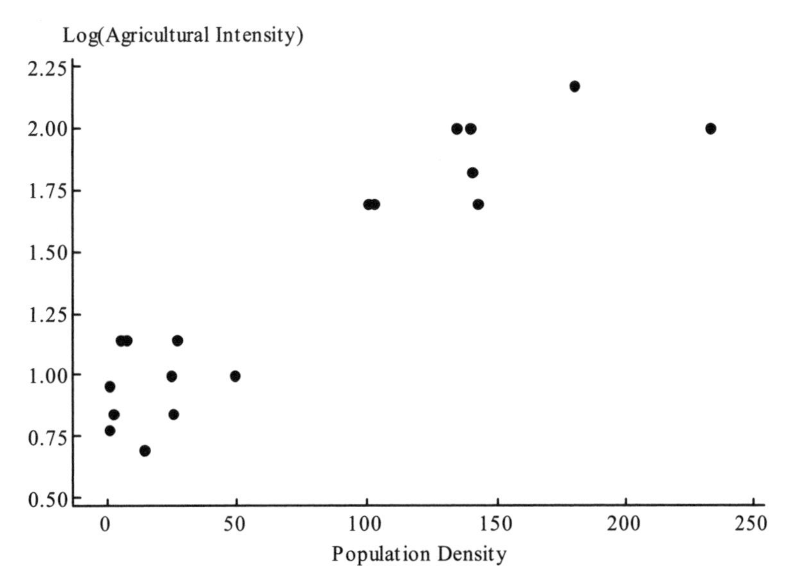

This transformation seems to have been successful in straightening the plot. Also, unlike the plot in Part (b), the variability of the quantity measured on the vertical axis does not seem to increase as x increases.

d

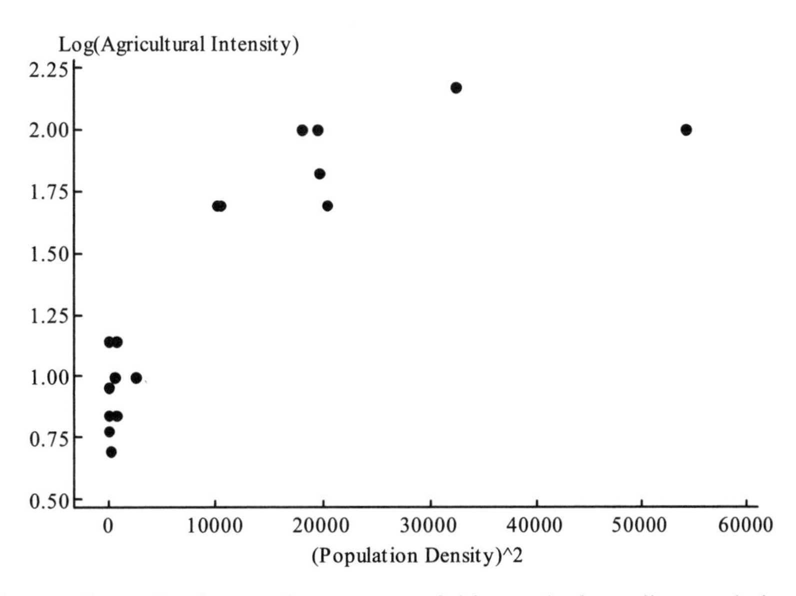

No, this transformation has not been successful in producing a linear relationship. There is a clear curve in the plot.

Chapter 6
Probability

6.1 A chance experiment is any activity or situation in which there is uncertainty about which of two or more possible outcomes will result. For example, a random number generator is used to select a whole number between 1 and 4, inclusive.

6.3 **a** {AA, AM, MA, MM}

b

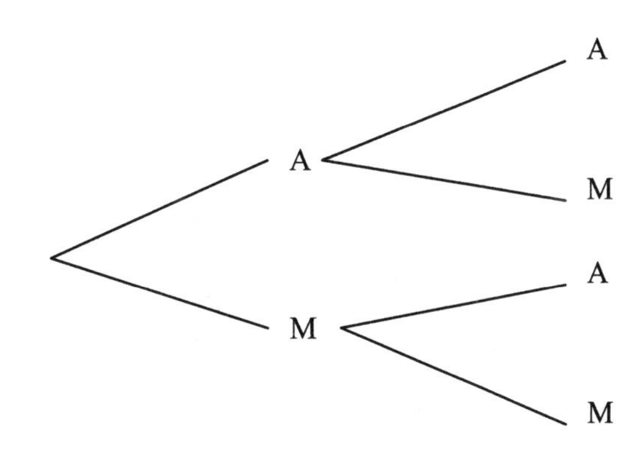

c **i** B = {AA, AM, MA}
ii C = {AM, MA}
iii D = {MM}. D is a simple event.

d *B and C* = {AM, MA}
B or C = {AA, AM, MA}

6.5 **a**

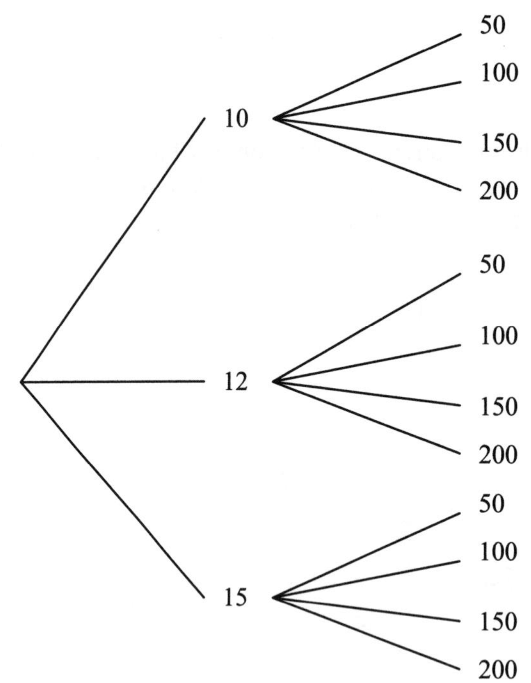

b $A^C = \{(15, 50), (15, 100), (15, 150), (15, 200)\}$

$$A \cup B = \begin{Bmatrix} (10, 50), (10, 100), (10, 150), (10, 200), (12, 50) \\ (12, 100), (12, 150), (12, 200), (15, 50), (15, 100) \end{Bmatrix}$$

$$A \cap B = \{(10, 50), (10, 100), (12, 50), (12, 100)\}$$

c *A* and *C* are not disjoint events.
B and *C* are disjoint events.

6.7 **a**

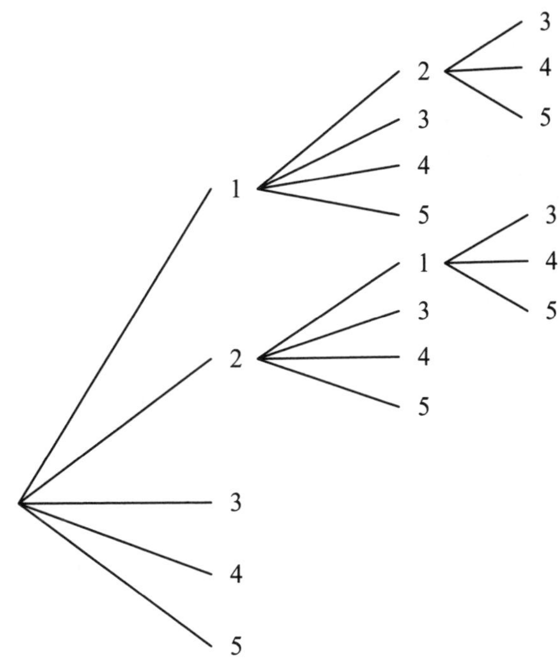

b $A = \{(3), (4), (5)\}$

c $C = \{1\ 2\ 5,\ 1\ 5,\ 2\ 1\ 5,\ 2\ 5,\ 5\}$

6.9 **a** $A = \{\text{NN, DNN}\}$

b $B = \{\text{DDNN}\}$

c There are an infinite number of outcomes.

6.11 **a**

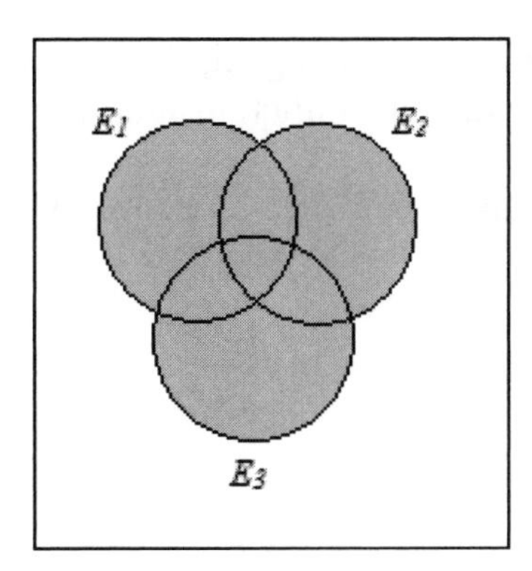

b

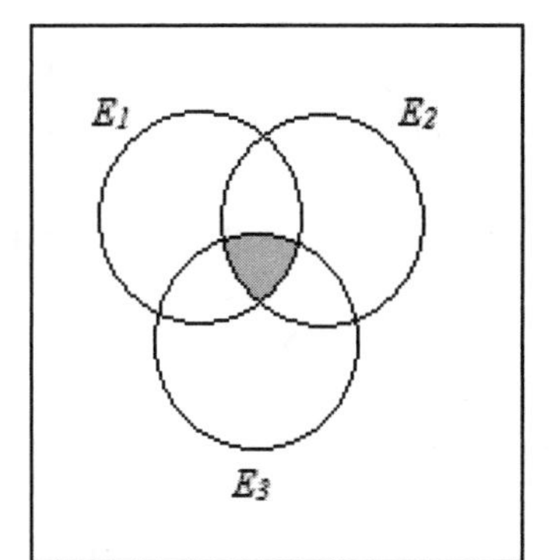

c

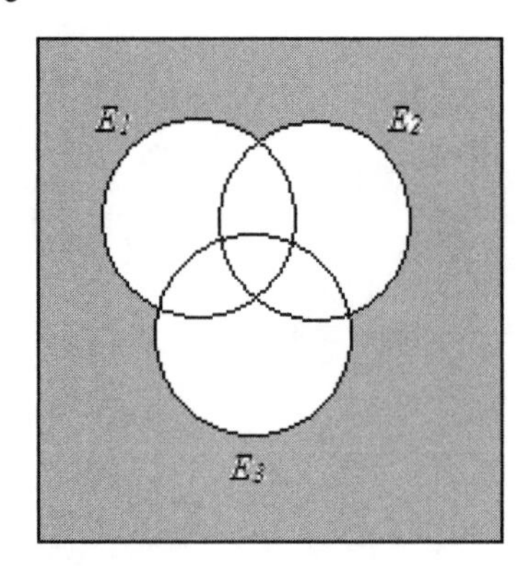

e

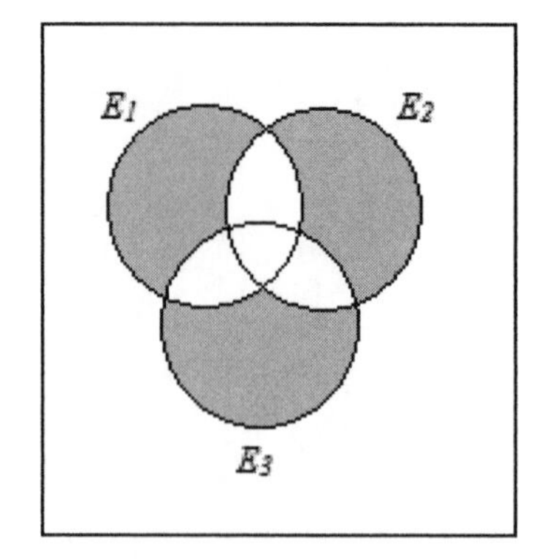

d

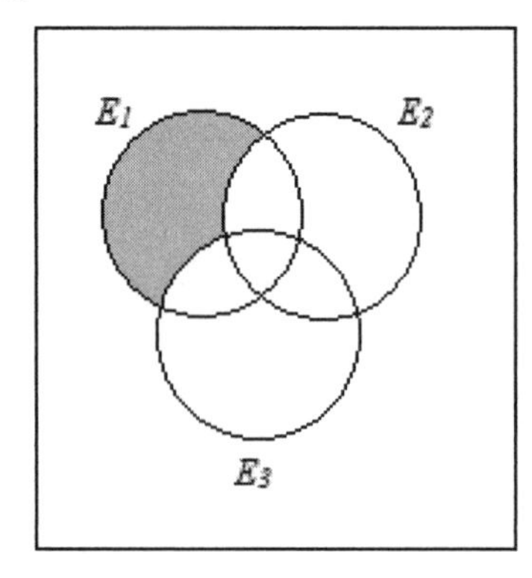

f

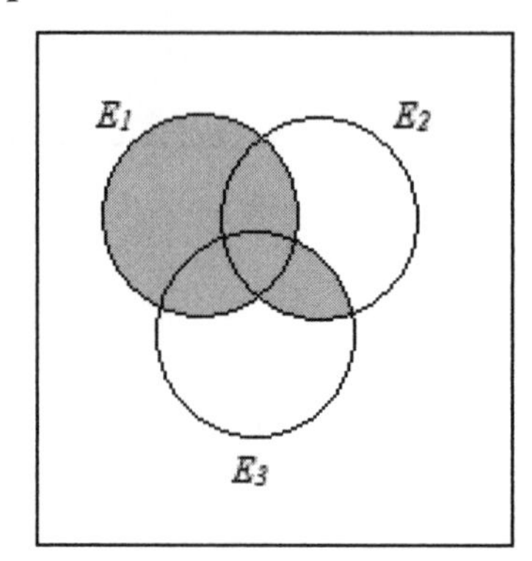

6.13 **a** The sample space is {expedited overnight delivery, expedited second business day delivery, standard delivery, delivery to the nearest store for customer pick-up}.

b **i** 1 − 0.1 − 0.3 − 0.4 = **0.2**
ii 0.1 + 0.3 = **0.4**
iii 0.4 + 0.2 = **0.6**

6.15 **a** The sample space is {fiction hardcover, fiction paperback, fiction digital, fiction audio, nonfiction hardcover, nonfiction paperback, nonfiction digital, nonfiction audio}

b No. For example, a customer is more likely to buy a paperback than a hardcover book.

c 0.15 + 0.08 + 0.45 + 0.04 = **0.72**

d 0.1 + 0.02 + 0.1 + 0.06 = **0.28**; 1 − 0.72 = **0.28**

e 0.15 + 0.45 + 0.1 + 0.1 = **0.8**

6.17 **a** There are 1000 different possible outcomes (000–999) and just one of those is the number 911. Thus the probability of the sequence 9-1-1 is 1/1000 = **0.001**.

b Classical. We used the fact that there are 1000 possible outcomes in the experiment, one of which consists of the sequence 9-1-1 being selected.

6.19 Assuming that the probability of dying in an accident on any one flight is 1/11,000,000, we can then conclude that *on average* a passenger will make 11,000,000 flights (which is one flight a day for 11,000,000/365 = 30,137 years) before dying in a crash (that is, assuming that a person could live that long). However, some passengers would survive more than this number of flights, and some fewer.

6.21 **a** Thirty-five percent of all tennis rackets purchased have a grip size of 4½ inches.

b $P(\text{not } 4\tfrac{1}{2} \text{ inches}) = 1 - 0.35 = \mathbf{0.65}$.

c $P(B) = 0.2 + 0.15 + 0.2 = \mathbf{0.55}$. Fifty-five percent of all tennis rackets purchased have oversize heads.

d $P(\text{grip size is at least } 4\tfrac{1}{2} \text{ inches}) = 0.2 + 0.15 + 0.15 + 0.2 = \mathbf{0.7}$.

6.23 **a** $P(\text{Bach or Beethoven or Brahms}) = 0.05 + 0.26 + 0.09 = \mathbf{0.4}$.

b $P(\text{not one of the S's}) = 1 - 0.12 - 0.07 = \mathbf{0.81}$.

c $P(\text{composer who wrote at least one symphony}) = 1 - 0.05 - 0.01 = \mathbf{0.94}$.

6.25 **a** $P(\text{just spades}) = 1287/2598960 = \mathbf{0.000495}$.
Since there are 1287 hands consisting entirely of spades, there are also 1287 hands consisting entirely of clubs, 1287 hands consisting entirely of diamonds, and 1287 hands consisting entirely of hearts. So the number of possible hands consisting of just one suit is 4(1287) = 5148. Thus, $P(\text{single suit}) = 5148/2598960 = \mathbf{0.00198}$.

b $P(\text{entirely spades and clubs with both suits represented}) = 63206/2598960 = \mathbf{0.024}$.

c The two suits could be spades and clubs, spades and diamonds, spades and hearts, clubs and diamonds, clubs and hearts, or diamonds and hearts. So there are six different combinations of two suits, with 63,206 possible hands for each combination. Therefore, $P(\text{exactly two suits}) = 6(63206)/2598960 = \mathbf{0.146}$.

6.27 **a** BC, BM, BP, BS, CM, CP, CS, MP, MS, PS

b 1/10 = **0.1**

c 4/10 = **0.4**

d There are three outcomes in which both representatives come from laboratory science subjects: BC, BP, and CP. Thus $P(\text{both from laboratory science}) = 3/10 = \mathbf{0.3}$.

6.29 **a** From $p + 2p + p + 2p + p + 2p = 1$ we see that $9p = 1$, and so $p = 1/9$. Therefore, $P(O_1) = P(O_3) = P(O_5) = \mathbf{1/9}$ and $P(O_2) = P(O_4) = P(O_6) = \mathbf{2/9}$.

b $P(\text{odd}) = P(O_1) + P(O_3) + P(O_5) = 1/9 + 1/9 + 1/9 = \mathbf{1/3}$.

$P(\text{at most 3}) = P(O_1) + P(O_2) + P(O_3) = 1/9 + 2/9 + 1/9 = \mathbf{4/9}$.

c From $c + 2c + 3c + 4c + 5c + 6c = 1$ we see that $21c = 1$, and so $c = 1/21$.
Thus $P(O_1) = \mathbf{1/21}$, $P(O_2) = \mathbf{2/21}$, $P(O_3) = 3/21 = \mathbf{1/7}$, $P(O_4) = \mathbf{4/21}$,
$P(O_5) = \mathbf{5/21}$, $P(O_6) = 6/21 = \mathbf{2/7}$. Therefore,
$P(\text{odd}) = P(O_1) + P(O_3) + P(O_5) = 1/21 + 3/21 + 5/21 = \mathbf{9/21}$ and
$P(\text{at most 3}) = P(O_1) + P(O_2) + P(O_3) = 1/21 + 2/21 + 3/21 = 6/21 = \mathbf{2/7}$.

6.31 **a** $P(\text{tattoo}) = 24/100 = \mathbf{0.24}$

b $P(\text{tattoo} \mid \text{age 18-29}) = 18/50 = \mathbf{0.36}$

c $P(\text{tattoo} \mid \text{age 30-50}) = 6/50 = \mathbf{0.12}$

d $P(\text{age 18-29} \mid \text{tattoo}) = 18/24 = \mathbf{0.75}$

6.33 **a** **0.72**

b The value 0.45 is the conditional probability that the selected individual drinks 2 or more cups a day given that he or she drinks coffee. We know this because the percentages given in the display add to 100, and yet we know that only 72% of Americans drink coffee. So the percentages given in the table must be the proportions *of coffee drinkers* who drink the given amounts.

6.35 **a** **i** If a person has been out of work for 1 month then the probability that the person will find work within the next month is 0.3.

ii If a person has been out of work for 6 months then the probability that the person will find work within the next month is 0.19.

b

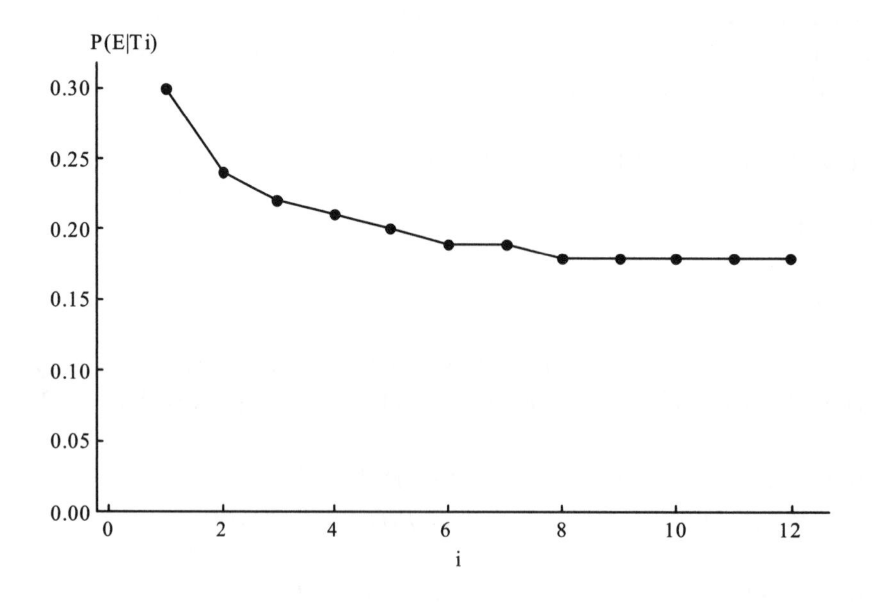

c The longer you have been unemployed the less likely you are to find a job during the next month, with the amount by which the likelihood decreases becoming smaller as the number of months out of work increases. It seems from the information given that, after a time, the probability of finding a job during the next month stabilizes at about 0.18.

6.37 $P(A \mid B)$ is larger. $P(A \mid B)$ is the probability that a randomly chosen professional basketball player is over six feet tall – a reasonably large probability. $P(B \mid A)$ is the probability that a randomly chosen person over six feet tall is a professional basketball player – a very small probability.

6.39 **a** $P(\text{former smoker}) = 99/294 = \mathbf{0.337}$.

b $P(\text{very harmful}) = 224/294 = \mathbf{0.762}$.

c $P(\text{very harmful} \mid \text{current smoker}) = 60/96 = \mathbf{0.625}$.

d $P(\text{very harmful} \mid \text{former smoker}) = 78/99 = \mathbf{0.788}$.

e $P(\text{very harmful} \mid \text{never smoked}) = 86/99 = \mathbf{0.869}$.

f Of the three smoking statuses, current smokers are the least likely to believe that smoking is very harmful and those who have never smoked are the most likely to think that smoking is very harmful. This is not surprising since you would expect those who smoke to be the most confident about the health prospects of a smoker, with those who formally smoked being a little more concerned, and those who have never smoked being the most concerned.

6.41 **a**

Age	Does Not Use Regularly	Uses Regularly	Total
18-24	0.09833	0.06833	**0.16667**
25-34	0.12167	0.04500	**0.16667**
35-44	0.12333	0.04333	**0.16667**
45-54	0.11667	0.05000	**0.16667**
55-64	0.11667	0.05000	**0.16667**
65 and older	0.13667	0.03000	**0.16667**
Total	**0.71333**	**0.28667**	**1.00000**

i $P(A_1) = \mathbf{0.167}$

ii $P(A_1 \cap S) = \mathbf{0.068}$

iii $P(A_1 \mid S) = P(A_1 \cap S)/P(S) = 0.06833/0.28667 = \mathbf{0.238}$

iv $P(\textit{not}\, A_1) = 1 - P(A_1) = 1 - 0.16667 = \mathbf{0.833}$

v $P(S \mid A_1) = P(S \cap A_1)/P(A_1) = 0.06833/0.16667 = \mathbf{0.41}$

vi $P(S \mid A_6) = P(S \cap A_6)/P(A_6) = 0.03/0.16667 = \mathbf{0.18}$

b The conditional probabilities tell us that 18–24-year-olds are more likely than seniors to regularly wear seat belts.

6.43 **a** Eighty-five percent of all calls to the station are for medial assistance.

b 0.15

c $(0.85)(0.85) = \mathbf{0.7225}$

d $(0.85)(0.15) = \mathbf{0.1275}$

e $P(\text{exactly one for medical assistance})$
$= P(\text{1st for medical} \cap \text{2nd not for medical}) + P(\text{1st not for medical} \cap \text{2nd for medical})$
$= (0.85)(0.15) + (0.15)(0.85) = \mathbf{0.255}$.

f It would seem reasonable to assume that the outcomes of successive calls (for medical assistance or not) do not affect each other. However, it is likely that at certain times of the day calls are more likely to be for medical assistance than at other times of the day. Therefore, if a call is chosen at random and found to be for medical assistance, then it becomes more likely that this call was received at one of these times. The next call, being at roughly the same time of the day, then has a more than 0.85 probability of being for medical assistance. Therefore it is not reasonable to assume that the outcomes of successive calls are independent.

6.45 $P(L) \cdot P(F) = (0.58)(0.5) = 0.29 \neq P(L \cap F)$. Therefore the events L and F are not independent.

6.47 No. The events are not independent because the probability of experiencing pain daily given that the person is male is not equal to the probability of experiencing pain daily given that the person is not male.

6.49 **a** $(0.1)\,(0.1)\,(0.1) = \mathbf{0.001}$. We have to assume that she deals with the three errands independently.

b $P(\text{remembers at least one}) = 1 - P(\text{forgets them all}) = 1 - 0.001 = \mathbf{0.999}$.

c $P(\text{remembers 1st, forgets 2nd, forgets 3rd}) = (0.9)(0.1)(0.1) = \mathbf{0.009}$.

6.51 **a** $P(\text{1-2 subsystem works}) = (0.9)(0.9) = \mathbf{0.81}$.

b $P(\text{1-2 subsystem doesn't work}) = 1 - 0.81 = \mathbf{0.19}$.
$P(\text{3-4 subsystem doesn't work}) = \mathbf{0.19}$.

c $P(\text{system won't work}) = (0.19)(0.19) = \mathbf{0.0361}$.
$P(\text{system will work}) = 1 - 0.0361 = \mathbf{0.9639}$.

d $P(\text{system won't work}) = (0.19)(0.19)(0.19) = 0.006859$.
So $P(\text{system will work}) = 1 - 0.006859 = \mathbf{0.993141}$.

e The probability that one particular subsystem will work is now (0.9) (0.9) (0.9) = 0.729. So the probability that the subsystem won't work is $1 - 0.729 = 0.271$. Therefore the probability that neither of the two subsystems works (and so the system doesn't work) is (0.271) (0.271) = 0.073441. So the probability that the system works is $1 - 0.073441 = \mathbf{0.926559}$.

6.53 **a** The expert was assuming that there was a 1 in 12 chance of a valve being in any one of the 12 clock positions and that the positions of the two air valves were independent.

b Since the car's wheels are probably the same size, if one of the wheels happens to have its air valve in the same position as before then the other wheel is likely also to have its air valve in the same position as before. Thus the positions of the two air valves are *not* independent, and 1/144 is smaller than the correct probability.

6.55 **a** No. If the first board selected is defective then it is less likely that the second board is defective than if the first board had not been defective.

b $P(not\, E_1) = 1 - P(E_1) = 1 - 40/5000 = \mathbf{0.992}$.

c If the first board is defective then 39 of the remaining 4999 boards are defective. So $P(E_2 \mid E_1) = 39/4999 = \mathbf{0.00780}$. If the first board is not defective then 40 of the remaining 4999 boards are defective. So $P(E_2 \mid not\, E_1) = 40/4999 = \mathbf{0.00800}$. If the first board is defective then it is slightly less likely that the second board will be defective than if the first board is not defective.

d Yes. Since the two probabilities calculated in Part (c) are roughly equal, it would be reasonable to say that the events E_1 and E_2 are approximately independent.

6.57 **a** $P(\text{both correct}) = (50/800)(50/800) = \mathbf{0.00391}$.

b $P(\text{2nd is correct} \mid \text{1st is correct}) = P(\text{2nd is correct} \cap \text{1st is correct})/P(\text{1st is correct})$. So

$$P(\text{1st is correct} \cap \text{2nd is correct}) = P(\text{1st is correct}) \cdot P(\text{2nd is correct} \mid \text{1st is correct})$$
$$= (50/800)(49/799) = \mathbf{0.00383}.$$

This probability is slightly smaller than the one in Part (a).

6.59 **a** $6/10 = \mathbf{0.6}$

b $P(F \mid E) = \mathbf{5/9}$.

c $P(F \mid E) = P(F \cap E)/P(E)$.
So $P(E\, and\, F) = P(F \cap E) = P(E) \cdot P(F \mid E) = (6/10)(5/9) = \mathbf{1/3}$.

6.61 **a** $P(E \cup F) = P(E) + P(F) - P(E \cap F) = 0.4 + 0.3 - 0.15 = \mathbf{0.55}$.

b The Venn diagram is shown below.

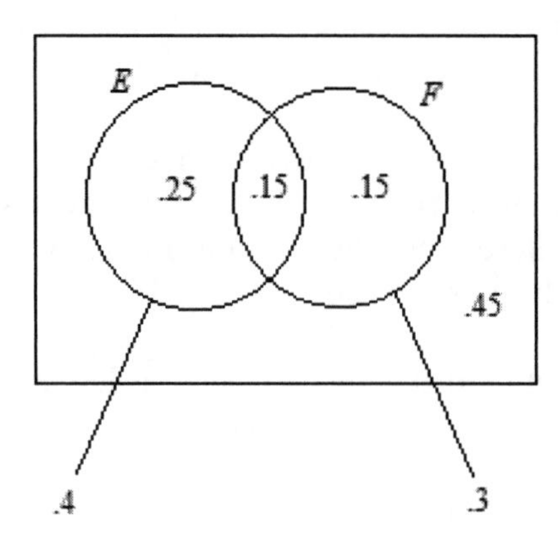

$P(\text{needn't stop at either light}) = 1 - P(E \cup F) = 1 - 0.55 = \mathbf{0.45}$.

c $P(\text{stops at exactly one}) = 0.25 + 0.15 = \mathbf{0.4}$.

d $P(\text{stops at just the first light}) = \mathbf{0.25}$.

6.63 **a** $P(F) = \mathbf{0.0303}$, $P(S \mid F) = \mathbf{0.7776}$, $P(D \mid S \cap F) = \mathbf{0.01012}$.

b $P(F \cap S \cap D) = P(F) \cdot P(S \mid F) \cdot P(D \mid S \cap F) = (0.0303)(0.7776)(0.0102) = \mathbf{0.000240}$.

6.65 **a** Yes. Since it is a *large* cable company, meaning that it has many customers, the outcome for the first customer (cable TV or not) will have little effect on the outcome for the second customer.

b $P(C_1 \cap C_2) = (0.8)(0.8) = \mathbf{0.64}$.

6.67 The statement is unlikely to be correct, since there is probably a large intersection between the set of ethnic minority students and the set of students who receive need-based financial aid.

6.69 **a** **i** $P(T) = \mathbf{0.307}$.

ii $P(T^C) = \mathbf{0.693}$.

iii $P(C \mid T) = \mathbf{0.399}$.

iv $P(L \mid T) = \mathbf{0.275}$.

v $P(C \cap T) = (0.307)(0.399) = \mathbf{0.122}$.

b 30.7% of faculty members use Twitter.
69.3% of faculty members do not use Twitter.
39.9% of faculty members who use Twitter also use it to communicate with students.
27.5% of faculty members who use Twitter also use it as a learning tool in the classroom.
12.2% of faculty members use Twitter and use it to communicate with students.

c By the law of total probability, $P(C) = P(C \cap T) + P(C \cap T^C)$. However, faculty members who do not use Twitter cannot possibly use it to communicate with students. Therefore $P(C \cap T^C) = 0$. Thus $P(C) = P(C \cap T) = \mathbf{0.122}$.

d Since faculty members who use Twitter as a learning tool must use Twitter, $P(L) = P(L \cap T) = P(T) \cdot P(L \mid T) = (0.307)(0.275) = \mathbf{0.084}$.

6.71 The reason that $P(C)$ is not the average of the three conditional probabilities is that there are different numbers of people driving the three different types of vehicle (and also that there are some drivers who are driving vehicles not included in those three types).

6.73 **a** **i** **0.99**
ii **0.01**
iii **0.99**
iv **0.01**

b
$$P(TD) = P(TD \cap C) + P(TD \cap D)$$
$$= P(C) \cdot P(TD \mid C) + P(D) \cdot P(TD \mid D)$$
$$= (0.99)(0.01) + (0.01)(0.99) = \mathbf{0.0198}.$$

c $P(C \mid TD) = P(C \cap TD)/P(TD) = (0.99)(0.01)/0.0198 = \mathbf{0.5}$.
Yes. The quote states that half of the dirty tests are false. The above probability states that half of the dirty tests are on people who are in fact clean. This confirms the statement in the quote.

6.75 **a** Yes. Eight percent of the large number of full-time workers are drug users, and 70% of the relatively small number of drug users are employed full-time.

b $P(D \mid E) = \mathbf{0.08}$; $P(E \mid D) = \mathbf{0.7}$.

c No. All we know is that $P(D \cap E)/P(E) = 0.08$ and that $P(D \cap E)/P(D) = 0.7$. This gives us two equations in three unknowns, and so we are unable to find any of the quantities involved. If we knew the percentage of the population who are employed full-time, then we would be able to find $P(D)$.

6.77 **a**

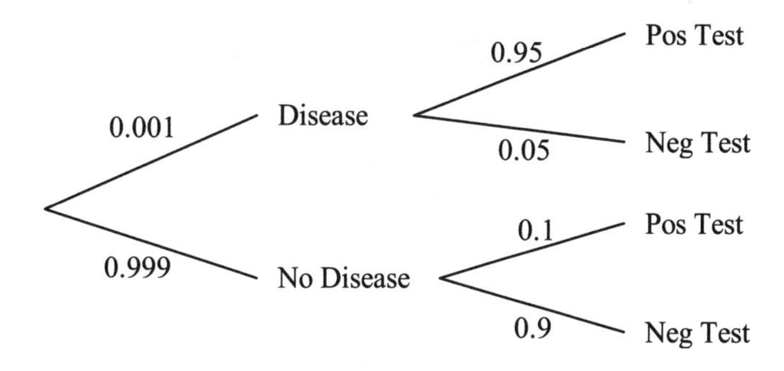

b P(has disease *and* positive test) $= (0.001)(0.95) = \mathbf{0.00095}$.

c P(positive test) $= (0.001)(0.95) + (0.999)(0.1) = \mathbf{0.10085}$.

d P(has disease | positive test) $= P$(has disease *and* positive test)$/P$(positive test)

$$= 0.00095/0.10085 = \mathbf{0.00942}.$$

This means that in less than 1% of positive tests the person actually has the disease. This comes about because so many more people do not have the disease than have the disease, and the minority of people who do not have the disease but still test positive greatly outnumber the people who *have* the disease and test positive.

6.79 **a**

	High School GPA			
Probation	**2.5 to <3.0**	**3.0 to <3.5**	**3.5 or Above**	**Total**
Yes	0.10	0.11	0.06	**0.27**
No	0.09	0.27	0.37	**0.73**
Total	**0.19**	**0.38**	**0.43**	**1.00**

b **0.27**

c **0.43**

d $P(\text{3.5 or above} \cap \text{probation}) = 0.06$. $P(\text{3.5 or above}) \cdot P(\text{probation}) = (0.43)(0.27) = 0.1161$. Since $P(\text{3.5 or above} \cap \text{probation}) \neq P(\text{3.5 or above}) \cdot P(\text{probation})$, the two outcomes are not independent.

e $P(\text{probation} \mid \text{2.5 to 3.0}) = 0.1/0.19 = \mathbf{0.526}$.

f $P(\text{probation} \mid \text{3.5 or above}) = 0.06/0.43 = \mathbf{0.140}$.

6.81

	Caucasian	**Hispanic**	**Black**	**Asian**	**American Indian**	**Total**
Monterey	163,000	139,000	24,000	39,000	4,000	**369,000**
San Luis Obispo	180,000	37,000	7,000	9,000	3,000	**236,000**
Santa Barbara	230,000	121,000	12,000	24,000	5,000	**392,000**
Ventura	430,000	231,000	18,000	50,000	7,000	**736,000**
Total	**1,003,000**	**528,000**	**61,000**	**122,000**	**19,000**	**1,733,000**

a 736/1733 = **0.425**

b 231/736 = **0.314**

c 231/528 = **0.4375**

d 9/1733 = **0.005**

e (122 + 180 + 37 + 7 + 3)/1733 = 349/1733 = **0.201**

f (39 + 24 + 50 + 180 + 37 + 7 + 3)/1733 = 340/1733 = **0.196**

g (1003/1733) (1003/1733) = **0.335**. [Note: It could be argued that this calculation should be (1003000/1733000)(1002999/1732999). However, since the given calculation yields a result that is very close to the result of this second calculation, and since the population figures given are clearly approximations, the given calculation will suffice.]

h Proportion who are not Caucasian $= 1 - 1003/1733 = 730/1733$. Therefore, when two people are selected at random, $P(\text{both are Caucasian}) = (730/1733)(730/1733) = \mathbf{0.177}$.

i (1003/1733)(730/1733) + (730/1733)(1003/1733) = **0.488**

j (369/1733)(369/1733) + (236/1733)(236/1733) + (392/1733)(392/1733) + (736/1733)(736/1733) = **0.295**

k $P(\text{same ethnic group}) = (1003/1733)(1003/1733) + (528/1733)(528/1733)$
$+ (61/1733)(61/1733) + (122/1733)(122/1733) + (19/1733)(19/1733) = 0.434.$
So $P(\text{different ethnic groups}) = 1 - 0.434 = \mathbf{0.566}$.

6.83 **a** The simulation could be designed as follows. Number the 20 companies/individuals applying for licenses 01–20, with the applications for 3 licenses coming from 01–06, the applications for 2 licenses coming from 07–15, and the applications for single licenses coming from 16–20. Assume that the individual with whom this question is concerned is numbered 16.

One run of the simulation is conducted as follows. Use two consecutive digits from a random number table to select a number between 01 and 20. (Ignore pairs of digits from the random number table that give numbers outside this range.) The number selected indicates the company/individual whose request is granted. Repeat the number selection, ignoring repeated numbers, until there are no licenses left. Make a note of whether or not individual 16 was awarded a license.

Perform the above simulation a large number of times. The probability that this particular individual is awarded a license is approximated by
(Number of runs in which 16 is awarded a license)/(Total number of runs).

b This does not seem to be a fair way to distribute licenses, since any given company applying for multiple licenses has roughly the same chance of obtaining all of its licenses as an individual applying for a single license has of obtaining his/her license. It might be fairer for companies who require two licenses to submit two separate applications and companies who require three licenses to submit three separate applications (with individuals/companies applying for single licenses submitting applications as before). Then 10 of the applications would be randomly selected and licenses would be awarded accordingly.

6.85 Results of the simulation will vary. The correct probability that the project is completed on time is 0.8468.

6.87 **a**

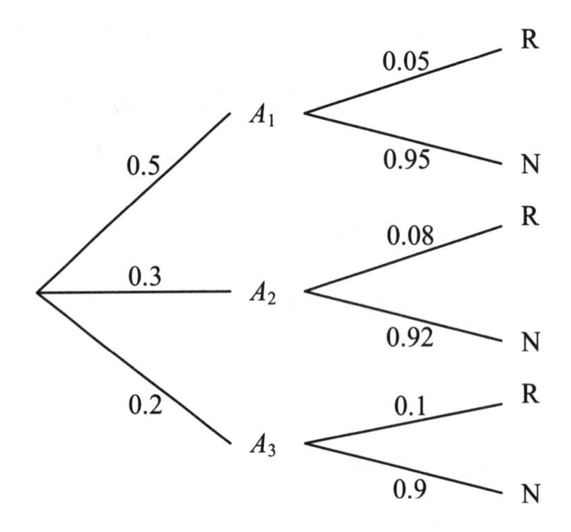

b $P(A_1 \cap \text{R}) = (0.5)(0.05) = \mathbf{0.025}$.

c $P(\text{R}) = (0.5)(0.05) + (0.3)(0.08) + (0.2)(0.1) = \mathbf{0.069}$.

6.89 **a** $P(C) = 27/193 = \mathbf{0.140}$. 14.0% of chat room users have criticized others.

b $P(O) = 42/193 = \mathbf{0.218}$. 21.8% of chat room users have been personally criticized.

c $P(C \cap O) = 19/193 = \mathbf{0.098}$. 9.8% of chat room users have criticized others and have been personally criticized.

d $P(C \mid O) = 19/42 = \mathbf{0.452}$. 45.2% of chat room users who have been personally criticized have criticized others.

e $P(O \mid C) = 19/27 = \mathbf{0.704}$. 70.4% of chat room users who have criticized others have been personally criticized.

6.91 They are dependent events, since someone who is attempting to quit is slightly more likely to return to smoking within two weeks if he/she does not use a nicotine aid than if he/she does use a nicotine aid.

6.93 **a**

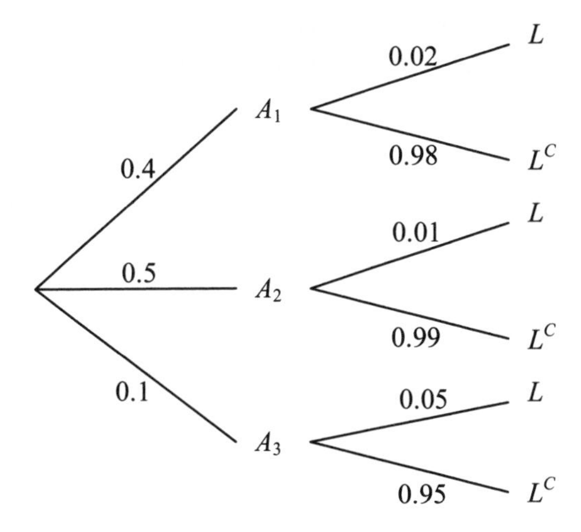

$P(L) = (0.4)(0.02) + (0.5)(0.01) + (0.1)(0.05) = \mathbf{0.018}.$

b $P(A_1 \mid L) = P(A_1 \cap L)/P(L) = (0.4)(0.02)/0.018 = \mathbf{0.444}.$
$P(A_2 \mid L) = P(A_2 \cap L)/P(L) = (0.5)(0.01)/0.018 = \mathbf{0.278}.$
$P(A_3 \mid L) = P(A_3 \cap L)/P(L) = (0.1)(0.05)/0.018 = \mathbf{0.278}.$

6.95 In Parts (a)–(c) below, examples of possible simulation plans are given.

a Use a single-digit random number to represent the outcome of the game. The digits 0–7 will represent a win for seed 1, and digits 8–9 will represent a win for seed 4.

b Use a single-digit random number to represent the outcome of the game. The digits 0–5 will represent a win for seed 2, and digits 6–9 will represent a win for seed 3.

c Use a single-digit random number to represent the outcome of the game.
If seed 1 won game 1 and seed 2 won game 2, the digits 0–5 will represent a win for seed 1, and digits 6–9 will represent a win for seed 2.
If seed 1 won game 1 and seed 3 won game 2, the digits 0–6 will represent a win for seed 1, and digits 7–9 will represent a win for seed 3.
If seed 4 won game 1 and seed 2 won game 2, the digits 0–6 will represent a win for seed 2, and digits 7–9 will represent a win for seed 4.
If seed 4 won game 1 and seed 3 won game 2, the digits 0–5 will represent a win for seed 3, and digits 6–9 will represent a win for seed 4.

d Answers will vary.

e Answers will vary.

f Answers will vary.

g The estimated probabilities from Parts (e) and (f) will differ because they are based on different sets of simulations. The estimate from Part (f) is likely to be the better one, since it is based on more runs of the simulation than the estimate from Part (e).

6.97

Individuals Chosen (by Years of Experience)	Total Number of Years' Experience	At least 15 years?
3, 6	9	No
3, 7	10	No
3, 10	13	No
3, 14	17	Yes
6, 7	13	No
6, 10	16	Yes
6, 14	20	Yes
7, 10	17	Yes
7, 14	21	Yes
10, 14	24	Yes

$P(\text{at least 15 years' experience}) = 6/10 = \mathbf{0.6}$.

6.99 **a** $(179 + 87)/2517 = \mathbf{0.106}$

b $(420 + 323 + 179 + 114 + 87)/2517 = \mathbf{0.446}$

c $1-(600+196+205+139)/2517 = \mathbf{0.547}$

6.101 **a** $P(E) = 20/25 = \mathbf{4/5}$.

b $P(F \mid E) = \mathbf{19/24}$.

c $P(G \mid E \cap F) = \mathbf{18/23}$.

d $P(\text{all good}) = (20/25)(19/24)(18/23) = \mathbf{0.496}$.

6.103 **a** $(0.8)(0.8)(0.8) = \mathbf{0.512}$

b The possible sequences and their probabilities are shown in the table below.

Sent by Transmitter	Sent by Relay 1	Sent by Relay 2	Sent by Relay 3	Probability
1	0	0	1	(0.2)(0.8)(0.2)
1	0	1	1	(0.2)(0.2)(0.8)
1	1	0	1	(0.8)(0.2)(0.2)
1	1	1	1	(0.8)(0.8)(0.8)

$$P(\text{1 is received}) = (0.2)(0.8)(0.2) + (0.2)(0.2)(0.8) + (0.8)(0.2)(0.2) + (0.8)(0.8)(0.8) = \mathbf{0.608}.$$

Chapter 7
Random Variables and Probability Distributions

Note: In this chapter, numerical answers to questions involving the normal distribution were found using statistical tables. Students using calculators or computers will find that their answers differ slightly from those given.

7.1 **a** Discrete

b Continuous

c Discrete

d Discrete

e Continuous

7.3 The possible y values are the positive integers. For the possible outcomes and their y values, answers will vary.

7.5 The possible values of y are the real numbers between 0 and 100, inclusive. The random variable y is continuous.

7.7 **a** 3, 4, 5, 6, 7

b $-3, -2, -1, 0, 1, 2, 3$

c 0, 1, 2

d 0, 1

7.9 **a** $p(4) = 1 - 0.65 - 0.2 - 0.1 - 0.04 = \mathbf{0.01}$.

b Over a large number of cartons, 20% will contain 1 broken egg.

c $P(y \le 2) = 0.65 + 0.2 + 0.1 = \mathbf{0.95}$. Over a large number of cartons, 95% will contain at most 2 broken eggs.

d $P(y < 2) = 0.65 + 0.2 = \mathbf{0.85}$. This probability is smaller than the one in Part (c) since it does not include the possibility of 2 broken eggs.

e $P(\text{exactly 10 unbroken}) = P(\text{exactly 2 broken}) = \mathbf{0.1}$.

f $P(\ge 10 \text{ unbroken}) = P(\le 2 \text{ broken}) = \mathbf{0.95}$.

7.11 **a** The probability that everyone who shows up can be accommodated is $P(x \le 100) = 0.05 + 0.1 + 0.12 + 0.14 + 0.24 + 0.17 = \mathbf{0.82}$.

b $1-0.82=\mathbf{0.18}$.

c For the person who is number 1 on the standby list to get a place on the flight, 99 or fewer people must turn up for the flight. The probability that this happens is $P(x \le 99)=0.05+0.1+0.12+0.14+0.24=\mathbf{0.65}$.

For the person who is number 3 on the standby list to get a place on the flight, 97 or fewer people must turn up for the flight. The probability that this happens is $P(x \le 99)=0.05+0.1+0.12=\mathbf{0.27}$.

7.13 Results of the simulation will vary.

7.15 **a** The sixteen possible outcomes, their probabilities, and the associated values of x, are shown in the table below.

Outcome	Probability	x
SSSS	$(0.2)(0.2)(0.2)(0.2)=0.0016$	4
SSSF	$(0.2)(0.2)(0.2)(0.8)=0.0064$	3
SSFS	$(0.2)(0.2)(0.8)(0.2)=0.0064$	3
SSFF	$(0.2)(0.2)(0.8)(0.8)=0.0256$	2
SFSS	$(0.2)(0.8)(0.2)(0.2)=0.0064$	3
SFSF	$(0.2)(0.8)(0.2)(0.8)=0.0256$	2
SFFS	$(0.2)(0.8)(0.8)(0.2)=0.0256$	2
SFFF	$(0.2)(0.8)(0.8)(0.8)=0.1024$	1
FSSS	$(0.8)(0.2)(0.2)(0.2)=0.0064$	3
FSSF	$(0.8)(0.2)(0.2)(0.8)=0.0256$	2
FSFS	$(0.8)(0.2)(0.8)(0.2)=0.0256$	2
FSFF	$(0.8)(0.2)(0.8)(0.8)=0.1024$	1
FFSS	$(0.8)(0.8)(0.2)(0.2)=0.0256$	2
FFSF	$(0.8)(0.8)(0.2)(0.8)=0.1024$	1
FFFS	$(0.8)(0.8)(0.8)(0.2)=0.1024$	1
FFFF	$(0.8)(0.8)(0.8)(0.8)=0.4096$	0

The probability distribution of x is given in the table below.

x	0	1	2	3	4
$p(x)$	0.4096	0.4096	0.1536	0.0256	0.0016

b There are two most likely values of x: **0 and 1**.

c $P(x \ge 2)=0.1536+0.0256+0.0016=\mathbf{0.1808}$.

7.17 **a** The smallest possible y value is **1**, and the corresponding outcome is **S**. The second smallest y value is **2**, and the corresponding outcome is **FS**.

b The set of positive integers

c $P(y=1)=P(\text{S})=\mathbf{0.7}$

$P(y=2)=P(\text{FS})=(0.3)(0.7)=\mathbf{0.21}$.

$P(y=3)=P(\text{FFS})=(0.3)(0.3)(0.7)=\mathbf{0.063}$.

$P(y=4) = P(\text{FFFS}) = (0.3)(0.3)(0.3)(0.7) = \mathbf{0.0189}$.
$P(y=5) = P(\text{FFFFS}) = (0.3)(0.3)(0.3)(0.3)(0.7) = \mathbf{0.00567}$.
The formula is $p(y) = (0.3)^{y-1}(0.7)$, for $y = 1, 2, 3, \ldots$

7.19

		1st Magazine			
		W	T	F	S
2nd Magazine	W	$y=0$ prob = 0.16	$y=1$ prob = 0.12	$y=2$ prob = 0.08	$y=3$ prob = 0.04
	T	$y=1$ prob = 0.12	$y=1$ prob = 0.09	$y=2$ prob = 0.06	$y=3$ prob = 0.03
	F	$y=2$ prob = 0.08	$y=2$ prob = 0.06	$y=2$ prob = 0.04	$y=3$ prob = 0.02
	S	$y=3$ prob = 0.04	$y=3$ prob = 0.03	$y=3$ prob = 0.02	$y=3$ prob = 0.01

The probability distribution of y is shown below.

y	0	1	2	3
$p(y)$	0.16	0.33	0.32	0.19

7.21 **a** $P(x \le 5) = (1/10)(5-0) = \mathbf{0.5}$.

b $P(3 \le x \le 5) = (1/10)(5-3) = \mathbf{0.2}$.

7.23 **a**

b The area of the rectangular region under the curve must be 1. The width of the rectangle is $20 - 7.5 = 12.5$. So, if the height is h, $12.5h = 1$. Therefore, $h = 1/12.5 = \mathbf{0.08}$.

c $P(x \le 12) = (0.08)(12 - 7.5) = \mathbf{0.36}$.

d $P(10 \le x \le 15) = (0.08)(15 - 10) = \mathbf{0.4}$.
$P(12 \le x \le 17) = (0.08)(17 - 12) = \mathbf{0.4}$.
These probabilities are equal because they are represented on the graph by rectangles of equal width and equal height.

7.25 **a** $P(x < 10) = 0.05(10 - 0) = \mathbf{0.5}$.
$P(x > 15) = 0.05(20 - 15) = \mathbf{0.25}$.

b $P(7 < x < 12) = 0.05(12 - 7) = \mathbf{0.25}$.

c $0.05(c - 0) = 0.9$, so $c = 0.9/0.05 = \mathbf{18}$.

7.27 **a** The density values at the relevant *x* values are given in the table below.

x	0	0.25	0.5	0.75	1
Density	0.5	0.75	1	1.25	1.5

$P(x \le 0.5) = (0.5 - 0)((0.5 + 1)/2) = \mathbf{0.375}$.
$P(0.25 < x < 0.5) = (0.5 - 0.25)((0.75 + 1)/2) = \mathbf{0.21875}$.
$P(x \ge 0.75) = (1 - 0.75)((1.25 + 1.5)/2) = \mathbf{0.34375}$.

b $\sigma_x = \sqrt{\dfrac{11}{144}} = 0.27639$. So $\mu_x - \sigma_x = 7/12 - 0.27639 = 0.30695$, and $\mu_x + \sigma_x = 7/12 + 0.27639 = 0.85972$. At $x = 0.30695$, density = 0.80695 and at $x = 0.85972$, density = 1.35972. So the probability that *x* is *less* than 1 standard deviation from its mean is $(0.85972 - 0.30695)((0.80695 + 1.35972)/2) = 0.599$. Therefore, the probability that *x* is *more* than 1 standard deviation from its mean is $1 - 0.599 = \mathbf{0.401}$.

7.29 **a** $\mu_y = 0(0.65) + 1(0.2) + 2(0.1) + 3(0.04) + 4(0.01) = \mathbf{0.56}$. Over a large number of cartons, the mean number of broken eggs per carton will be 0.56.

b $P(y < \mu_y) = P(y < 0.56) = P(y = 0) = 0.65$. In **65%** of cartons, the number of broken eggs will be less than μ_y. This result is greater than 0.5 because the distribution of *y* is positively skewed.

c The mean is not $(0 + 1 + 2 + 3 + 4)/5$ since, for example, far more cartons have 0 broken eggs than 4 broken eggs, and so 0 needs a much greater weighting than 4 in the calculation of the mean.

7.31 **a** 0.02 + 0.03 + 0.09 = **0.14**.

b $\mu - 2\sigma = 4.66 - 2(1.2) = 2.26$. $\mu + 2\sigma = 4.66 + 2(1.2) = 7.06$. The values of *x* more than 2 standard deviations away from the mean are **1** and **2**. The probability that *x* is more than 2 standard deviations away from its mean is 0.02 + 0.03 = **0.05**.

7.33 **a**

d_1	0	650	2000
$p(d_1)$	1/3	1/3	1/3

b $E(d_1) = 0(1/3) + 650(1/3) + 2000(1/3) = \mathbf{883.333}$.

c $\sigma_{d_1}^2 = (0 - 883.333)^2(1/3) + (650 - 883.333)^2(1/3) + (2000 - 883.333)^2(1/3) = 693888.889$.

So $\sigma_{d_1} = \sqrt{693888.889} = \mathbf{833.000}$.

d The probability distributions of d_2 and d_3 are shown below.

d_2	0	650	1350
$p(d_2)$	1/3	1/3	1/3

d_3	0	1350	2000
$p(d_3)$	1/3	1/3	1/3

Neither of these probability distributions is the same as the probability distribution of d_1.

e The possible values of t are calculated in the table below.

Where meeting held	d_1	d_2	t	Probability
Columbus	0	650	650	1/3
Des Moines	650	0	650	1/3
Boise	2000	1350	3350	1/3

This tells us that the probability distribution of t is as shown below.

t	650	3350
$p(t)$	2/3	1/3

f The probability distribution of d_2 is given below.

d_2	0	650	1350
$p(d_2)$	1/3	1/3	1/3

From this we get $E(d_2) = 666.667$ and $\sigma_{d_2} = 551.261$.

From the probability distribution of t we get $E(t) = 1550$ and $\sigma_t = 1272.792$.

i $E(d_1) + E(d_2) = 883.333 + 666.667 = 1550 = E(t)$, so the statement is true. (Note that the statement *has* to be true, since $t = d_1 + d_2$, and since for any two random variables x and y, regardless of whether or not they are independent, $E(x + y) = E(x) + E(y)$.)

ii $\sigma_{d_1}^2 + \sigma_{d_2}^2 = 883.000^2 + 551.261^2 = 303888.889 + 693888.889 = 997777.778$, but $\sigma_t^2 = 1272.792^2 = 1620000$. So the statement is false.

7.35 **a** $E(x) = 15(0.1) + 30(0.3) + 60(0.6) = \mathbf{46.5}$ seconds.

b The probability distribution of y is shown below.

y	500	800	1000
$p(y)$	0.1	0.3	0.6

$E(y) = 500(0.1) + 800(0.3) + 1000(0.6) = 890$. The average amount paid is **\$890**.

7.37 For the first distribution shown below, $\mu_x = 3$ and $\sigma_x = 1.414$, while for the second distribution $\mu_x = 3$ and $\sigma_x = 1$.

x	1	2	3	4	5
$p(x)$	0.2	0.2	0.2	0.2	0.2

x	1	2	3	4	5
$p(x)$	0.1	0.1	0.6	0.1	0.1

7.39 **a** Whether y is positive or negative tells us whether or not the peg will fit into the hole.

b $\mu_y = \mu_{x_2} - \mu_{x_1} = 0.253 - 0.25 = \mathbf{0.003}$.

c $\sigma_y = \sqrt{\sigma_{x_1}^2 + \sigma_{x_2}^2} = \sqrt{(0.006)^2 + (0.002)^2} = \mathbf{0.00632}$.

d Yes. Since we have no reason to believe that the pegs are being specially selected to match the holes (or vice versa), it is reasonable to think that x_1 and x_2 are independent.

e Since 0 is less than half a standard deviation from the mean in the distribution of y, it is relatively likely that a value of y will be negative, and therefore that the peg will be too big to fit the hole.

7.41 **a** $\mu_x = 0(0.05) + 1(0.1) + \cdots + 5(0.1) = \mathbf{2.8}$.
$\sigma_x^2 = (0 - 2.8)^2(0.5) + \cdots + (5 - 2.8)^2(0.1) = 1.66$.
So $\sigma_x = \sqrt{1.66} = \mathbf{1.288}$.

b $\mu_y = 0(0.5) + 1(0.3) + 2(0.2) = \mathbf{0.7}$.
$\sigma_y^2 = (0 - 0.7)^2(0.5) + (1 - 0.7)^2(0.3) + (2 - 0.7)^2(0.2) = 0.61$.
So $\sigma_y = \sqrt{0.61} = \mathbf{0.781}$.

c The amount of money collected in tolls from cars is (number of cars)(\$3) = $3x$.
$\mu_{3x} = 3\mu_x = 3(2.8) = \mathbf{8.4}$.
$\sigma_{3x}^2 = 3^2\sigma_x^2 = 9(1.66) = \mathbf{14.94}$.

d The amount of money collected in tolls from buses is (number of buses)(\$10) = $10y$.
$\mu_{10y} = 10\mu_y = 10(0.7) = \mathbf{7}$.
$\sigma_{10y}^2 = 10^2\sigma_y^2 = 100(0.61) = \mathbf{61}$.

e $\mu_z = \mu_{x+y} = \mu_x + \mu_y = 2.8 + 0.7 = \mathbf{3.5}$.
$\sigma_z^2 = \sigma_{x+y}^2 = \sigma_x^2 + \sigma_y^2 = 1.66 + 0.61 = \mathbf{2.27}$.

f $\mu_w = \mu_{3x+10y} = \mu_{3x} + \mu_{10y} = 8.4 + 7 = \mathbf{15.4}$.

$\sigma_w^2 = \sigma_{3x+10y}^2 = \sigma_{3x}^2 + \sigma_{10y}^2 = 14.94 + 61 = \mathbf{75.94}$.

7.43 **a** **6** outcomes have exactly one success. The outcomes are: SFFFFF, FSFFFF, FFSFFF, FFFSFF, FFFFSF, FFFFFS.

b The number of outcomes with exactly 10 successes is $20!/(10!10!) = \mathbf{184756}$.
The number of outcomes with exactly 15 successes is $20!/(15!5!) = \mathbf{15504}$.
The number of outcomes with exactly 5 successes is $20!/(5!15!) = \mathbf{15504}$.

7.45 **a** $p(4) = \left(6!/(4!2!)\right)(0.8)^4(0.2)^2 = \mathbf{0.246}$. Over a large number of random selections of 6 passengers, 24.6% will have exactly four people resting or sleeping.

b $p(6) = (0.8)^6 = \mathbf{0.262}$.

c $P(x \geq 4) = p(4) + p(5) + p(6) = \left(6!/(4!2!)\right)(0.8)^4(0.2)^2 + \left(6!/(5!1!)\right)(0.8)^5(0.2)^1 + (0.8)^6$
$= \mathbf{0.901}$.

7.47 **a** $p(2) = \left(5!/(2!3!)\right)(0.25)^2(0.75)^3 = \mathbf{0.264}$.

b $P(x \leq 1) = p(0) + p(1) = 0.23730 + 0.39551 = \mathbf{0.633}$.

c $P(2 \leq x) = P(x \geq 2) = 1 - P(x \leq 1) = 1 - 0.633 = \mathbf{0.367}$.

d $P(x \neq 2) = 1 - P(x = 2) = 1 - 0.264 = \mathbf{0.736}$.

7.49 **a** Using Appendix Table 9, $P(>15 \text{ have firewall}) = p(16) + p(17) + p(18) + p(19) + p(20)$
$= 0.218 + 0.205 + 0.137 + 0.058 + 0.012 = \mathbf{0.630}$.

b Using Appendix Table 9, $P(>15 \text{ have firewall}) = p(16) + p(17) + p(18) + p(19) + p(20)$
$= 0.000 + 0.000 + 0.000 + 0.000 + 0.000 = 0.000$. The required probability is **0.000** when rounded to the nearest one-thousandth.

c If the true proportion of computer owners who have a firewall installed is 0.4 then $p(14) = \left(20!/(14!6!)\right)(0.4)^{14}(0.6)^6 = 0.005$. If the true proportion of computer owners who have a firewall installed is 0.8 then $p(14) = \left(20!/(14!6!)\right)(0.8)^{14}(0.2)^6 = 0.109$. So if, in a random sample of 20 computer owners, 14 are observed to have a firewall installed, it is more likely that the true proportion of computer owners who have a firewall installed is **0.8**.

7.51 **a** $P(\text{all 10 pass}) = (0.85)^{10} = \mathbf{0.197}$.

b $P(>2 \text{ fail}) = 1 - P(\leq 2 \text{ fail}) = 1 - \left(P(0 \text{ fail}) + P(1 \text{ fails}) + P(2 \text{ fail})\right)$

$$=1-\left((0.85)^{10}+\frac{10!}{1!9!}(0.15)(0.85)^{9}+\frac{10!}{2!8!}(0.15)^{2}(0.85)^{8}\right)=\mathbf{0.180}.$$

c $\mu_x = 500(0.15) = \mathbf{75}$. $\sigma_x = \sqrt{500(0.15)(0.85)} = \mathbf{7.984}$.

d The value 25 is more than 6 standard deviations below the mean in the distribution of x. It is therefore surprising that fewer than 25 are said to have failed the test.

7.53 Expected number showing damage $= 2000(0.1) = \mathbf{200}$.
Standard deviation $= \sqrt{2000(0.1)(0.9)} = \mathbf{13.416}$.

7.55 **a** Binomial distribution with $n = 100$ and $p = 0.2$

b Expected score $= 100(0.2) = \mathbf{20}$.

c $\sigma_x^2 = 100(0.2)(0.8) = \mathbf{16}$, $\sigma_x = \sqrt{16} = \mathbf{4}$.

d A score of 50 is $(50-20)/4 = 7.5$ standard deviations from the mean in the distribution of x. So a score of over 50 is very unlikely.

7.57 **a** Using Appendix Table 9, $P(\text{program is implemented}) = P(x \le 15) = 0.012 + 0.004 + 0.001$ $= \mathbf{0.017}$.

b Using Appendix Table 9, if $p = 0.7$, $P(\text{program is implemented}) = P(x \le 15)$ $= 0.091 + 0.054 + 0.027 + 0.011 + 0.004 + 0.002 = 0.189$. So
$P(\text{program is not implemented}) = 1 - 0.189 = \mathbf{0.811}$.
Using Appendix Table 9, if $p = 0.6$, $P(\text{program is not implemented}) = P(x > 15)$
$= 0.151 + 0.120 + 0.080 + 0.045 + 0.020 + 0.007 + 0.002 = \mathbf{0.425}$.

c The error probability when $p = 0.7$ is now $0.811 + p(15) = 0.811 + 0.092 = \mathbf{0.903}$. The error probability when $p = 0.6$ is now $0.424 + p(15) = 0.424 + 0.161 = \mathbf{0.585}$.

7.59 **a** For a random variable to be binomially distributed, it must represent the number of "successes" in a fixed number of trials. This is not the case for the random variable x described.

b The distribution of x is geometric with $p = 0.08$.

i $p(4) = (0.92)^3(0.08) = \mathbf{0.062}$.

ii $P(x \le 4) = p(1) + p(2) + p(3) + p(4) = 0.08 + (0.92)(0.08) + (0.92)^2(0.08) + (0.92)^3(0.08)$ $= \mathbf{0.284}$.

iii $P(x > 4) = 1 - P(x \le 4) = 1 - 0.284 = \mathbf{0.716}$.

iv $P(x \ge 4) = p(4) + P(x > 4) = 0.06230 + 0.71639 = \mathbf{0.779}$.

c The process described is for songs to be randomly selected until a song by the particular artist is played.

i If the process were to be repeated many times, on 6.2% of occasions exactly four songs would be played up to and including the first song by this artist.
ii If the process were to be repeated many times, on 28.4% of occasions at most four songs would be played up to and including the first song by this artist.
iii If the process were to be repeated many times, on 71.6% of occasions more than four songs would be played up to and including the first song by this artist.
iv If the process were to be repeated many times, on 77.9% of occasions at least four songs would be played up to and including the first song by this artist.

The differences between the four probabilities lie in the underlined phrases.

7.61 **a** $P(x \le 2) = p(1) + p(2) = 0.05 + (0.95)(0.05) = \mathbf{0.0975}$.

b $p(4) = (0.95)^3(0.05) = \mathbf{0.043}$.

c $P(x > 4) = 1 - P(x \le 4) = 1 - \left((0.05) + (0.95)(0.05) + (0.95)^2(0.05) + (0.95)^3(0.05)\right) = \mathbf{0.815}$.

(Alternatively, note that for more than four boxes to be purchased, the first four boxes bought must not contain a prize. So $P(x > 4) = (0.95)^4 = \mathbf{0.815}$.)

7.63 **a** **0.9599**

b **0.2483**

c $1 - 0.8849 = \mathbf{0.1151}$.

d $1 - 0.0024 = \mathbf{0.9976}$.

e $0.7019 - 0.0132 = \mathbf{0.6887}$

f $0.8413 - 0.1587 = \mathbf{0.6826}$.

g **1.0000**

7.65 **a** **0.9909**

b **0.9909**

c **0.1093**

d $0.9996 - 0.8729 = \mathbf{0.1267}$

e $0.2912 - 0.2206 = \mathbf{0.0706}$

f $1 - 0.9772 = \mathbf{0.0228}$

g $1 - 0.0004 = \mathbf{0.9996}$

h **1.0000**

7.67 **a** **−1.96**

b **−2.33**

c **−1.645**

d $P(z < z^*) = 1 - 0.02 = 0.98$. So $z^* = \mathbf{2.05}$.

e $P(z < z^*) = 1 - 0.01 = 0.99$. So $z^* = \mathbf{2.33}$.

f $P(z > z^*) = 0.2/2 = 0.1$. So $P(z < z^*) = 1 - 0.1 = 0.9$. Therefore $z^* = \mathbf{1.28}$.

7.69 **a** $P(z > z^*) = 0.05/2 = 0.025$. So $P(z < z^*) = 1 - 0.025 = 0.975$. So $z^* = \mathbf{1.96}$.

b $P(z > z^*) = 0.1/2 = 0.05$. So $P(z < z^*) = 1 - 0.05 = 0.95$. So $z^* = \mathbf{1.645}$.

c $P(z > z^*) = 0.02/2 = 0.01$. So $P(z < z^*) = 1 - 0.01 = 0.99$. So $z^* = \mathbf{2.33}$.

d $P(z > z^*) = 0.08/2 = 0.04$. So $P(z < z^*) = 1 - 0.04 = 0.96$. So $z^* = \mathbf{1.75}$.

7.71 **a** $P(x < 5) = P\big(z < (5-5)/0.2\big) = P(z < 0) = \mathbf{0.5}$.

b $P(x < 5.4) = P\big(z < (5.4-5)/0.2\big) = P(z < 2) = \mathbf{0.9772}$.

c $P(x \le 5.4) = P(x < 5.4) = \mathbf{0.9772}$.

d $P(4.6 < x < 5.2) = P\big((4.6-5)/0.2 < z < (5.2-5)/0.2\big) = P(-2 < z < 1) = 0.8413 - 0.0228$
$= \mathbf{0.8185}$.

e $P(x > 4.5) = P\big(z > (4.5-5)/0.2\big) = P(z > -2.5) = 1 - P(z < -2.5) = 1 - 0.0062 = \mathbf{0.9938}$.

f $P(x > 4.0) = P\big(z > (4.0-5)/0.2\big) = P(z > -5) = 1 - P(z < -5) = 1 - 0.0000 = \mathbf{1.0000}$.

7.73 If $P(z > z^*) = 0.1$ then $P(z < z^*) = 0.9$; so $z^* = 1.28$. Thus $x = \mu + (z^*)\sigma = 1.6 + (1.28)(0.4)$ $= 2.113$. The worst 10% of vehicles are those with emission levels greater than **2.113** parts per billion.

7.75 **a** Let the left atrial diameter be x. $P(x < 24) = P\big(z < (24-26.4)/4.2\big) = P(z < -0.57) = \mathbf{0.2843}$.

b $P(x > 32) = P\big(z > (32-26.4)/4.2\big) = P(z > 1.33) = \mathbf{0.0918}$.

c $P(25 < x < 30) = P\big((25-26.4)/4.2 < z < (30-26.4)/4.2\big) = P(-0.33 < z < 0.86)$
$= 0.8051 - 0.3707 = \mathbf{0.4344}$.

d If $P(z > z^*) = 0.2$, then $P(z < z^*) = 0.8$; so $z^* = 0.84$. Thus $x = \mu + (z^*)\sigma = 26.4 + (0.84)(4.2)$ $= \mathbf{29.928}$ mm.

7.77 Let the carbon monoxide exposure be x.

$P(x > 20) = P\left(z > (20 - 18.6)/5.7\right) = P(z > 0.25) = 1 - P(z < 0.25) = 1 - 0.5987 = \mathbf{0.4013}$.

$P(x > 25) = P\left(z > (25 - 18.6)/5.7\right) = P(z > 1.12) = 1 - P(z < 1.12) = 1 - 0.8686 = \mathbf{0.1314}$.

7.79 Let the diameter of the cork produced be x.

$P(2.9 < x < 3.1) = P\left((2.9 - 3.05)/0.01 < z < (3.1 - 3.05)/0.01\right) = P(-15 < z < 5) = 1.0000$.

A cork made by the machine in this exercise is almost certain to meet the specifications. This machine is therefore preferable to the one in the Exercise 7.78.

7.81 The fastest 10% of applicants are those with the lowest 10% of times. If $P(z < z^*) = 0.1$, then $z^* = -1.28$. The corresponding time is $\mu + (z^*)\sigma = 120 + (-1.28)(20) = 94.4$. Those with times less than **94.4** seconds qualify for advanced training.

7.83 **a**

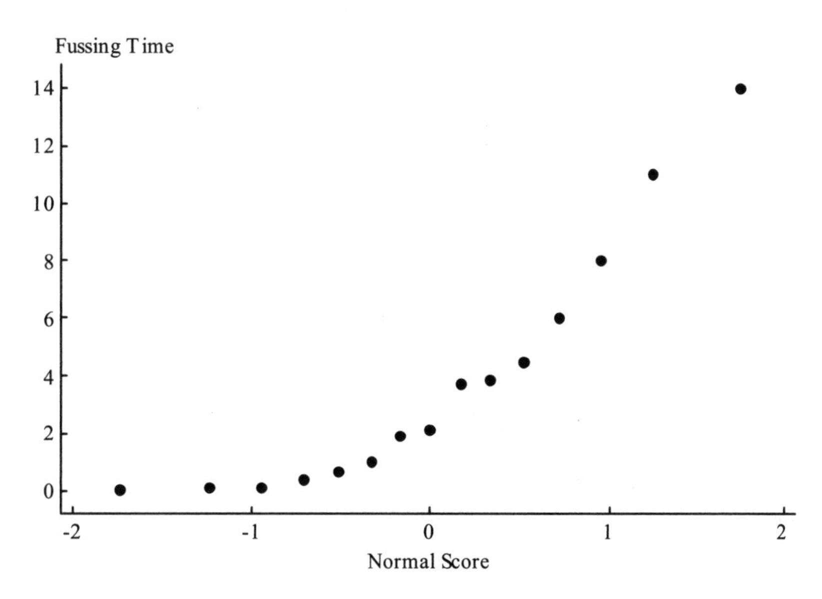

The clear curve in the normal probability plot tells us that the distribution of fussing times is not normal.

b The square roots of the data values are shown in the table below.

Fussing Time	Normal Score	sqrt(Fussing Time)
0.05	-1.739	0.22361
0.10	-1.245	0.31623
0.15	-0.946	0.38730
0.40	-0.714	0.63246
0.70	-0.515	0.83666
1.05	-0.333	1.02470
1.95	-0.165	1.39642
2.15	0.000	1.46629
3.70	0.165	1.92354
3.90	0.335	1.97484
4.50	0.515	2.12132
6.00	0.714	2.44949
8.00	0.946	2.82843
11.00	1.245	3.31662
14.00	1.739	3.74166

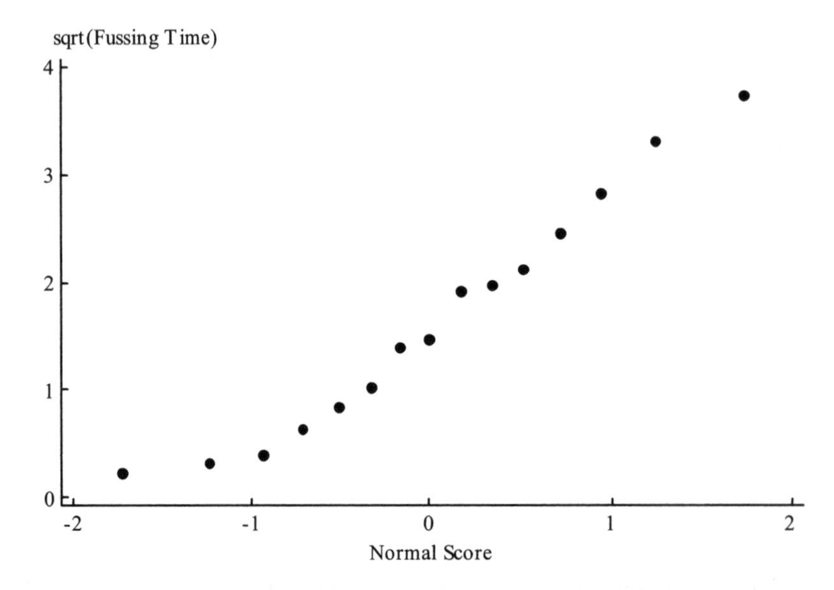

The transformation results in a pattern that is much closer to being linear than the pattern in Part (a).

7.85 a

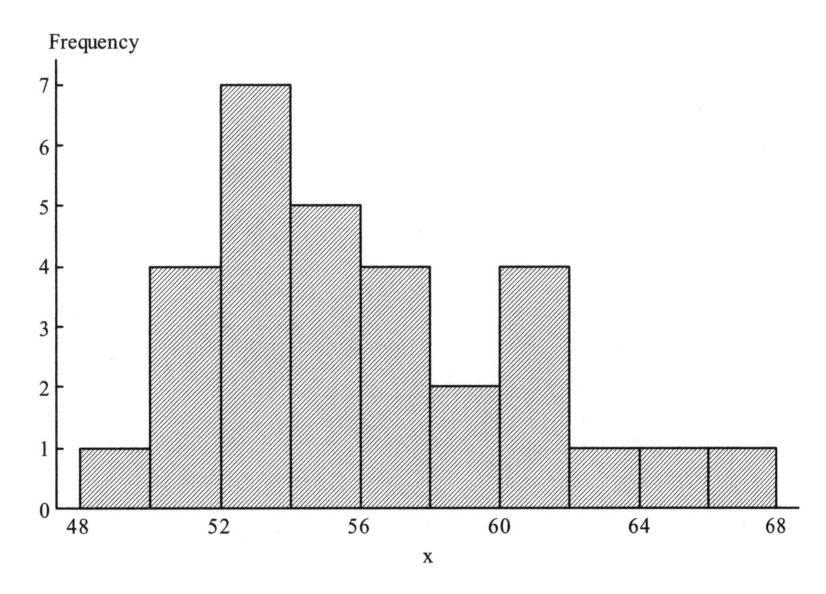

No. The distribution of x is positively skewed.

b

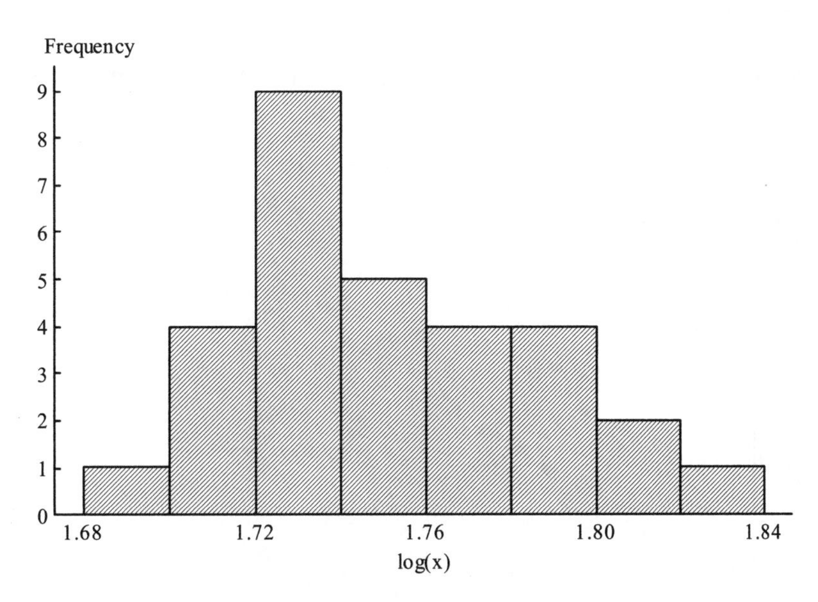

Yes. The histogram shows a distribution that is slightly closer to being symmetric than the distribution of the untransformed data.

c

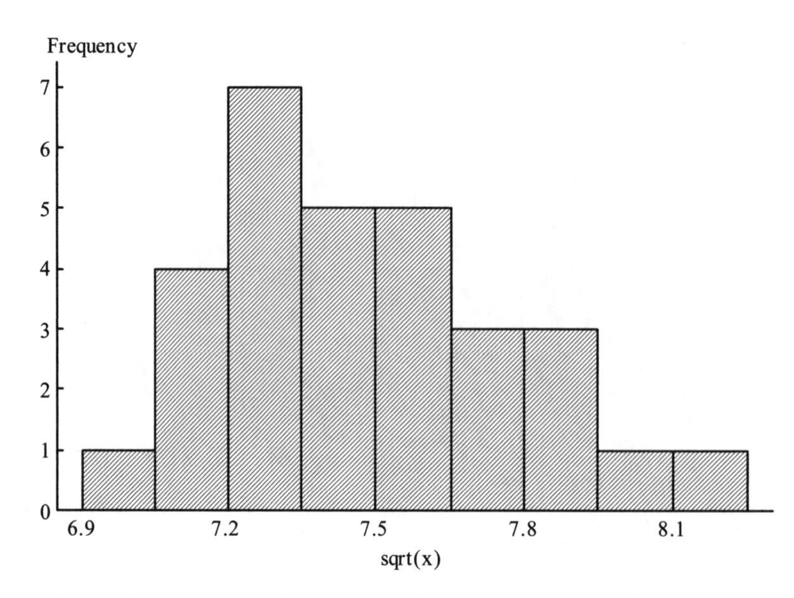

Both transformations produce histograms that are closer to being symmetric than the histogram of the untransformed data, but neither transformation produces a distribution that is truly close to being normal.

7.87 Yes. The curve in the normal probability plot suggests that the distribution is not normal.

7.89

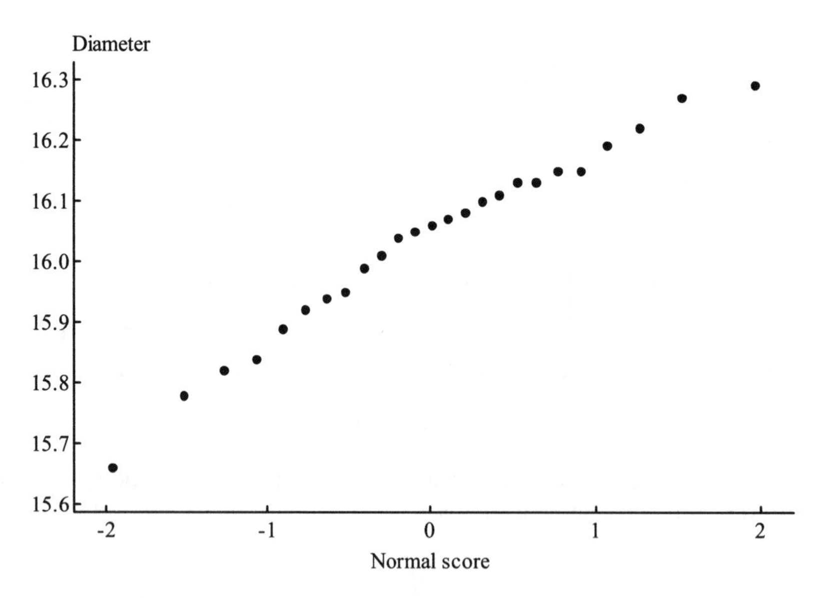

Since the pattern in the normal probability plot is very close to being linear, it is plausible that disk diameter is normally distributed.

7.91 a

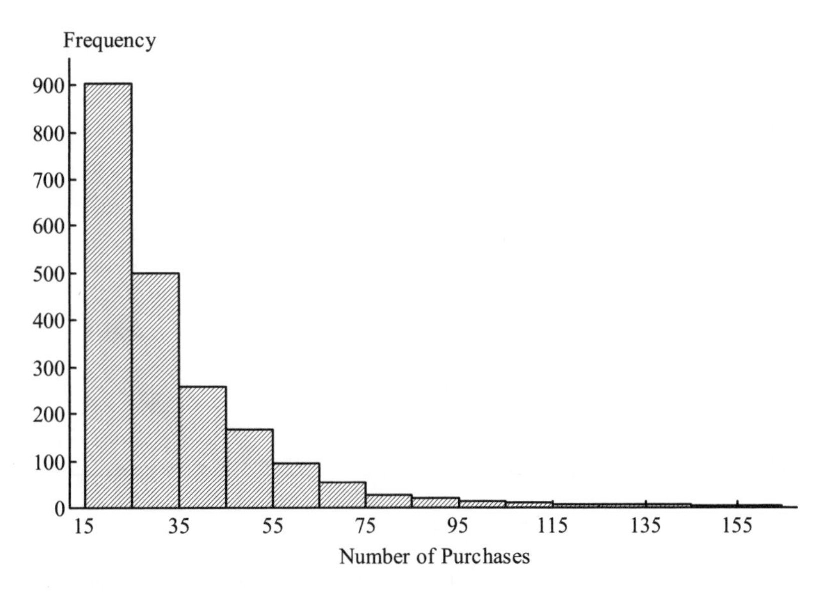

The histogram is positively skewed.

b

Interval	Frequency	Relative Frequency = Freq./2071	Interval Width	Density = Rel. Freq. / Int. Wdth.
√10 to <√20	904	0.437	1.310	0.333
√20 to <√30	500	0.241	1.005	0.240
√30 to <√40	258	0.125	0.847	0.147
√40 to <√50	167	0.081	0.747	0.108
√50 to <√60	94	0.045	0.675	0.067
√60 to <√70	56	0.027	0.621	0.044
√70 to <√80	26	0.013	0.578	0.022
√80 to <√90	20	0.010	0.543	0.018
√90 to <√100	13	0.006	0.513	0.012
√100 to <√110	9	0.004	0.488	0.009
√110 to <√120	7	0.003	0.466	0.007
√120 to <√130	6	0.003	0.447	0.006
√130 to <√140	6	0.003	0.430	0.007
√140 to <√150	3	0.001	0.415	0.003
√150 to <√160	0	0.000	0.402	0.000
√160 to <√170	2	0.001	0.389	0.002

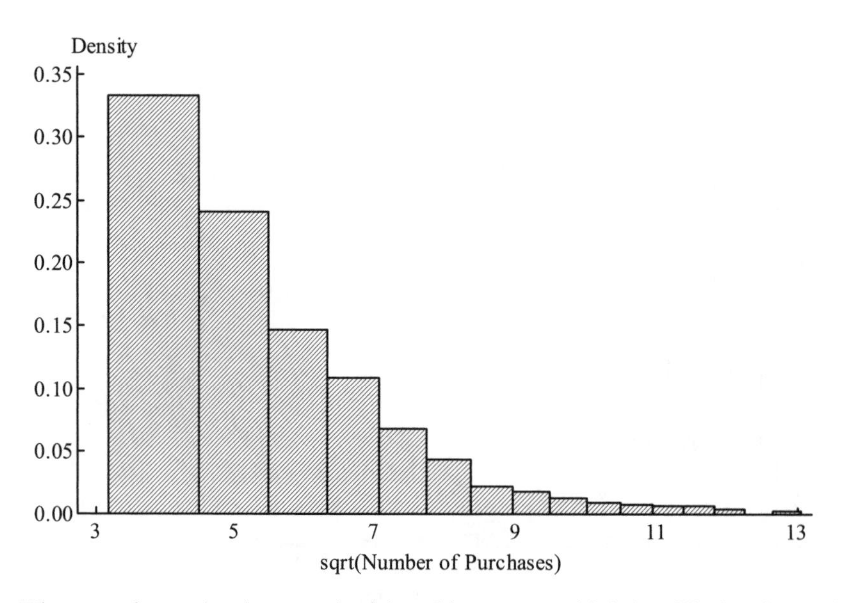

No. The transformation has resulted in a histogram which is still clearly positively skewed.

7.93 Yes. In each case the transformation has resulted in a histogram that is much closer to being symmetric than the original histogram.

7.95 **a** Let x = number of items.
$P(x \le 120) \approx P\big(z \le (120.5-150)/10\big) = P(z \le -2.95) = \mathbf{0.0016}$.

b $P(x \ge 125) \approx P\big(z \ge (124.5-150)/10\big) = P(z \ge -2.55) = 1 - P(z \le -2.55) = 1 - 0.0054 = \mathbf{0.9946}$.

c $P(135 \le x \le 160) \approx P\big((134.5-150)/10 \le z \le (160.5-150)/10\big) = P(-1.55 \le z \le 1.05)$
$= 0.8531 - 0.0606 = \mathbf{0.7925}$.

7.97 **a** $P(x = 30) \approx P\big((29.5-30)/3.4641 \le z \le (30.5-30)/3.4641\big) = P(-0.14 \le z \le 0.14)$
$= 0.5557 - 0.4443 = \mathbf{0.1114}$.

b $P(x = 25) \approx P\big((24.5-30)/3.4641 \le z \le (25.5-30)/3.4641\big) = P(-1.59 \le z \le -1.30)$
$= 0.0968 - 0.0559 = \mathbf{0.0409}$.

c $P(x \le 25) \approx P\big(z \le (25.5-30)/3.4641\big) = P(z \le -1.30) = \mathbf{0.0968}$.

d $P(25 \le x \le 40) \approx P\big((24.5-30)/3.4641 \le z \le (40.5-30)/3.4641\big) = P(-1.59 \le z \le 3.03)$
$= 0.9988 - 0.0559 = \mathbf{0.9429}$.

e $P(25 < x < 40) \approx P\big((25.5-30)/3.4641 \le z \le (39.5-30)/3.4641\big) = P(-1.30 \le z \le 2.74)$
$= 0.9969 - 0.0968 = \mathbf{0.9001}$.

7.99 **a** There is a fixed number of trials, where the probability of an undetected tumor is the same in each trial, and the outcomes of the trials are independent since the sample is random. Thus it is reasonable to think that the distribution of x is binomial.

b Yes, since $np = 500(0.031) = 15.5 \geq 10$ and $n(1-p) = 500(0.969) = 484.5 \geq 10$.

c $\mu = np = 500(0.031) = 15.5$ and $\sigma = \sqrt{np(1-p)} = \sqrt{500(0.031)(0.969)} = 3.8755$.

i $P(x < 10) \approx P\left(z \leq (9.5 - 15.5)/3.8755\right) = P(z \leq -1.55) = \mathbf{0.0606}$.

ii $P(10 \leq x \leq 25) \approx P\left((9.5 - 15.5)/3.8755 \leq z \leq (25.5 - 15.5)/3.8755\right) = P(-1.55 \leq z \leq 2.58)$ $= 0.9951 - 0.0606 = \mathbf{0.9345}$.

iii $P(x > 20) \approx P\left(z \geq (20.5 - 15.5)/3.8755\right) = P(z \geq 1.29) = 1 - P(z \leq 1.29) = 1 - 0.9015$ $= \mathbf{0.0985}$.

d **i** In a large number of random samples of 500 women diagnosed with cancer in one breast, approximately 6.06% will contain less than 10 women with an undetected tumor in the other breast.

ii In a large number of random samples of 500 women diagnosed with cancer in one breast, approximately 93.45% will contain between 10 and 25 (inclusive) women with an undetected tumor in the other breast.

iii In a large number of random samples of 500 women diagnosed with cancer in one breast, approximately 9.85% will contain more than 20 women with an undetected tumor in the other breast.

7.101 The normal approximation is reasonable since $np = 100(0.25) = 25 \geq 10$ and $n(1-p) = 100(0.75)$ $= 75 \geq 10$. Also, $\mu = np = 25$ and $\sigma = \sqrt{np(1-p)} = \sqrt{100(0.25)(0.75)} = 4.3301$.

a $P(20 \leq x \leq 30) \approx P\left((19.5 - 25)/4.3301 \leq z \leq (30.5 - 25)/4.3301\right) = P(-1.27 \leq z \leq 1.27)$ $= 0.8980 - 0.1010 = \mathbf{0.7970}$.

b $P(20 < x < 30) \approx P\left((20.5 - 25)/4.3301 \leq z \leq (29.5 - 25)/4.3301\right) = P(-1.04 \leq z \leq 1.04)$ $= 0.8508 - 0.1492 = \mathbf{0.7016}$.

c $P(x \geq 35) \approx P\left(z \geq (34.5 - 25)/4.3301\right) = P(z \geq 2.19) = 1 - P(z \leq 2.19) = 1 - 0.9857 = \mathbf{0.0143}$.

d $\mu - 2\sigma = 25 - 2(4.3301) = 16.3398$, and $\mu + 2\sigma = 25 + 2(4.3301) = 33.6602$.
So $P(x$ is less than two st. devs. from mean$) = P(16.3398 \leq x \leq 33.6602) = P(17 \leq x \leq 33)$ $\approx P\left((16.5 - 25)/4.3301 \leq z \leq (33.5 - 25)/4.3301\right) = P(-1.96 \leq z \leq 1.96) = 0.9750 - 0.0250$ $= 0.9500$. Therefore the probability that x is more than 2 standard deviations from its mean is $1 - 0.9500 = \mathbf{0.05}$.

7.103 **a** No, since $np = 50(0.05) = 2.5 < 10$.

b Now $n = 500$ and $p = 0.05$, so $np = 500(0.05) = 25 \geq 10$. So the techniques of this section can be used. Using $\mu = np = 25$ and $\sigma = \sqrt{np(1-p)} = \sqrt{500(0.05)(0.95)} = 4.8734$, $P(\text{at least 20 are defective}) \approx P\left(z \geq (19.5 - 25)/4.8734\right) = P(z \geq -1.13) = 1 - P(z \leq -1.13)$ $= 1 - 0.1292 = \mathbf{0.8708}$.

7.105 **a** No. The proportion of the population that is being sampled is 5000/40000 = 0.125, which is more than 5%.

b We have $n = 100$ and $p = 11000/40000 = 0.275$, so $\mu = np = 100(0.275) = \mathbf{27.5}$ and $\sigma = \sqrt{np(1-p)} = \sqrt{100(0.275)(0.725)} = \mathbf{4.465}$.

c No. Since n is being doubled, the standard deviation, which is $\sqrt{np(1-p)}$, is multiplied by $\sqrt{2}$.

7.107 **a** $P(x \le 3) = 0.1 + 0.15 + 0.2 + 0.25 = \mathbf{0.7}$.

b $P(x < 3) = 0.1 + 0.15 + 0.2 = \mathbf{0.45}$.

c $P(x \ge 3) = 1 - P(x < 3) = 1 - 0.45 = \mathbf{0.55}$.

d $P(2 \le x \le 5) = 0.2 + 0.25 + 0.2 + 0.06 = \mathbf{0.71}$.

e If 2 lines are not in use, then 4 lines are in use.
If 3 lines are not in use, then 3 lines are in use.
If 4 lines are not in use, then 2 lines are in use.
So the required probability is $P(2 \le x \le 4) = 0.2 + 0.25 + 0.2 = \mathbf{0.65}$.

f If 4 lines are not in use, then 2 lines are in use.
If 5 lines are not in use, then 1 line is in use.
If 6 lines are not in use, then 0 lines are in use.
So the required probability is $P(0 \le x \le 2) = 0.1 + 0.15 + 0.2 = \mathbf{0.45}$.

7.109 **a** $p(2) = (0.8)(0.8) = \mathbf{0.64}$.

b $p(3) = P(\text{UAA or AUA}) = (0.2)(0.8)(0.8) + (0.8)(0.2)(0.8) = \mathbf{0.256}$.

c For y to be 5, the fifth battery's voltage must be acceptable. The four outcomes are UUUAA, UUAUA, UAUUA, AUUUA. So $p(5) = 4(0.2)(0.2)(0.2)(0.8)(0.8) = \mathbf{0.02048}$.

d In order for it to take y selections to find two acceptable batteries, the first $y-1$ batteries must include exactly 1 acceptable battery (and therefore $y-2$ unacceptable batteries), and the yth battery must be acceptable. There are $y-1$ ways in which the first $y-1$ batteries can include exactly 1 acceptable battery. So $p(y) = (y-1)(0.8)^1(0.2)^{y-2} \cdot (0.8)$ $= (y-1)(0.2)^{y-2}(0.8)^2$.

7.111 Let the fuel efficiency of a randomly selected car of this type be x.

a $P(29 < x < 31) = P\big((29-30)/1.2 < z < (31-30)/1.2\big) = P\big(-0.83 < z < 0.83\big)$ $= 0.7967 - 0.2033 = \mathbf{0.5934}$.

b $P(x<25)=P\left(z<(25-30)/1.2\right)=P\left(z<-4.17\right)=0.0000$. Yes, you are very unlikely to select a car of this model with a fuel efficiency of less than 25 mpg.

c $P(x>32)=P\left(z>(32-30)/1.2\right)=P\left(z>1.67\right)=0.0475$. So, if 3 cars are selected, the probability that they all have fuel efficiencies more than 32 mpg is $(0.0475)^3=\mathbf{0.0001}$.

d If $P(z>z^*)=0.95$, then $z^*=-1.645$. So $c=\mu+(z^*)\sigma=30-(1.645)(1.2)=\mathbf{28.026}$.

7.113 $\sigma_x^2=(0-1.2)^2(0.54)+\cdots+(4-1.2)^2(0.2)=\mathbf{2.52}$.
$\sigma_x=\sqrt{2.52}=\mathbf{1.587}$.

7.115 Let the life of a randomly chosen battery be x.

a $P(x>4)=P\left(z>(4-6)/0.8\right)=P(z>-2.5)=0.9938$. For the player to function for at least 4 hours, both batteries have to last for at least 4 hours. The probability that this happens is $(0.9938)^2=\mathbf{0.9876}$.

b $P(x>7)=P\left(z>(7-6)/0.8\right)=P(z>1.25)=0.1056$. For the player to function for at least 7 hours, both batteries have to last for at least 7 hours. The probability that this happens is $(0.1056)^2=0.0112$. So the probability that the player works for at most 7 hours is $1-0.0112=\mathbf{0.9888}$.

c We need the probability that both batteries last longer than c hours to be 0.05. So for either one of the two batteries, we need the probability that it lasts longer than c hours to be $\sqrt{0.05}=0.2236$ (so that the probability that both batteries last longer than c hours is $(0.2236)^2=0.05$). Now, if $P(z>z^*)=0.2236$, then $z=0.76$. So $c=\mu+(z^*)\sigma$ $=6+(0.76)(0.8)=\mathbf{6.608}$.

7.117 **a** No, since 5 ft 7 in. is 67 inches, and if x = height of a randomly chosen women, then $P(x<67)=P\left(z<(67-66)/2\right)=P(z<0.5)=0.6915$, which is not more than 94%.

b About **69%** of women would be excluded by the height restriction.

7.119 **a** The possible combinations of arrival times and the corresponding values of w are given in the table below.

		Allison's arrival time					
		1	**2**	**3**	**4**	**5**	**6**
Terri's arrival time	**1**	0	1	2	3	4	5
	2	1	0	1	2	3	4
	3	2	1	0	1	2	3
	4	3	2	1	0	1	2
	5	4	3	2	1	0	1
	6	5	4	3	2	1	0

There is a total of 36 outcomes in the table, so the probability of a particular value of w is the number of occurrences of that value divided by 36. This tells us that the probability distribution of w is as shown in the table below.

w	0	1	2	3	4	5
$p(w)$	1/6	5/18	2/9	1/6	1/9	1/18

b $E(w) = 0(1/6) + 1(5/18) + 2(2/9) + 3(1/6) + 4(1/9) + 5(1/18) = \mathbf{1.944}$ hours.

7.121 **a** $p(4) = P(\text{KKKK}) + P(\text{LLLL}) = (0.4)^4 + (0.6)^4 = \mathbf{0.1552}$.

b $p(5) = P(\underbrace{\text{LLLK}}_{\text{in any order}}\text{L}) + P(\underbrace{\text{KKKL}}_{\text{in any order}}\text{K}) = 4(0.6)^3(0.4)(0.6) + 4(0.4)^3(0.6)(0.4) = \mathbf{0.2688}$.

c $p(6) = P(\underbrace{\text{LLLKK}}_{\text{in any order}}\text{L}) + P(\underbrace{\text{KKKLL}}_{\text{in any order}}\text{K}) = {}_5C_3(0.6)^3(0.4)^2(0.6) + {}_5C_3(0.4)^3(0.6)^2(0.4) = 0.29952.$

$p(7) = P(\underbrace{\text{LLLKKK}}_{\text{in any order}}\text{L}) + P(\underbrace{\text{LLLKKK}}_{\text{in any order}}\text{K}) = {}_6C_3(0.6)^3(0.4)^3(0.6) + {}_6C_3(0.6)^3(0.4)^3(0.4)$

$= 0.27648.$

So the probability distribution of x is as shown below.

x	4	5	6	7
$p(x)$	0.1552	0.2688	0.29952	0.27648

d $E(x) = 4(0.1552) + 5(0.2688) + 6(0.29952) + 7(0.27648) = \mathbf{5.697}$.

7.123 If $P(z \geq z^*) = 0.15$, then $z^* = 1.04$. So the lowest score to be designated an A is given by $\mu + (z^*)\sigma = 78 + (1.04)(7) = 85.28$. Since $89 > 85.28$, yes, I received an A.

7.125 If $P(z < z^*) = 0.2$, then $z = -0.84$. So the lowest 20% of lifetimes consist of those less than $\mu + (z^*)\sigma = 700 + (-0.84)(50) = 658$. So, if the bulbs are replaced every **658 hours**, then only 20% will have already burned out.

Cumulative Review Exercises

CR7.1 Obtain a group of volunteers (we'll assume that 60 people are available for this). Randomly assign the 60 people to two groups, A and B. (This can be done by writing the people's names on slips of paper, placing the slips in a hat, and drawing 30 slips at random. The people whose names are on those slips should be placed in Group A. The remaining people should be placed in Group B.) Meet with each person individually. For people in Group A offer an option of being given $5 or for a coin to be flipped. Tell the person that if the coin lands heads, he/she will be given $10, but if the coin lands tails, he/she will not be given any money. Note the person's choice, and then proceed according to the option the person has chosen. For people in Group B, give the person two $5 bills, and then offer a choice of returning one of the $5 bills, or flipping a coin. Tell the person that if the coin lands heads, he/she will keep both $5 bills, but if the coin lands tails, he/she must return both of the $5 bills. Note the person's choice, and then proceed according to the option the person has chosen. Once you have met with all the participants, compare the two groups in terms of the proportions choosing the gambling options.

CR7.3 No. The percentages given in the graph are said to be, for each year, the "percent increase in the number of communities installing" red-light cameras. This presumably means the percent increase in the number of communities with red-light cameras *installed*, in which case the positive results for all of the years 2003 to 2009 show that a great many more communities had red-light cameras installed in 2009 than in 2002.

CR7.5 We need $P(15.8 \le b_1 + b_2 \le 15.9)$. Since $\mu_{b_1} = \mu_{b_2} = 8$, and $\sigma_{b_1} = \sigma_{b_2} = 0.2$, $\mu_{b_1+b_2} = 8+8=16$ and $\sigma_{b_1+b_2} = \sqrt{(0.2)^2 + (0.2)^2} = 0.2828$.

So $P(15.8 \le b_1 + b_2 \le 15.9) = P\big((15.8-16)/0.2828 \le z \le (15.9-16)/0.2828\big)$

$= P(-0.71 \le z \le -0.35) = 0.3632 - 0.2389 = \mathbf{0.1243}$.

CR7.7 $P(\text{Service 1} \mid \text{Late}) = P(\text{Service 1} \cap \text{Late})/P(\text{Late}) = (0.3)(0.1)/\big((0.3)(0.1)+(0.7)(0.08)\big) = 0.349$.
So $P(\text{Service 2} \mid \text{Late}) = 1 - 0.349 = 0.651$. Thus **Service 2** is more likely to have been used.

CR7.9
1. $P(M) = 0.021$.
2. $P(M \mid B) = 0.043$.
3. $P(M \mid W) = 0.07$.

CR7.11

To say that a quantity x is 30% more than a quantity y means that the amount by which x exceeds y is 30% of y; that is, in mathematical notation, that $y - x = 0.3y$. Dividing both sides by 0.3, we see that this can be written as $(y-x)/y = 0.3$. Here we are told that drivers who live within one mile of a restaurant are 30% more likely to have an accident than those who do not live within one mile of a restaurant; in other words, that $P(A \mid R)$ is 30% more than $P(A \mid R^C)$. Thus statement **iv** is correct: $\big(P(A \mid R) - P(A \mid R^C)\big)/P(A \mid R^C) = 0.3$. None of the other statements is correct.

CR7.13

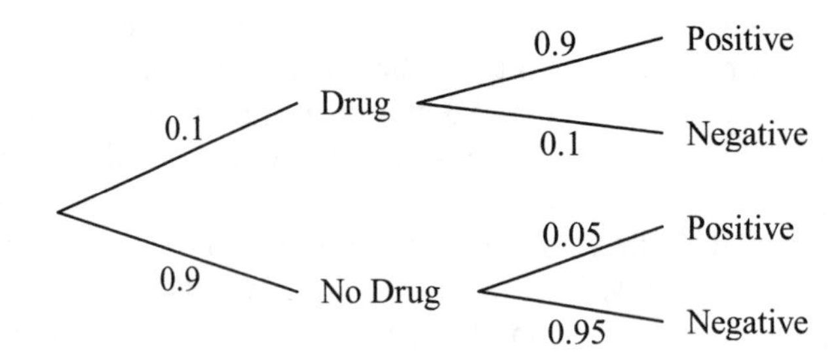

a $P(\text{Drug} \cap \text{Positive}) = (0.1)(0.9) = \mathbf{0.09}$.

b $P(\text{No Drug} \cap \text{Positive}) = (0.9)(0.05) = \mathbf{0.045}$.

c $P(\text{Positive}) = 0.09 + 0.045 = \mathbf{0.135}$.

d $P(\text{Drug} \mid \text{Positive}) = P(\text{Drug} \cap \text{Positive})/P(\text{Positive}) = 0.09/0.135 = \mathbf{0.667}$.

CR7.15

a $\mu_x = 1(0.2) + \cdots + 4(0.1) = \mathbf{2.3}$.

b $\sigma_x^2 = (1-2.3)^2(0.2) + \cdots + (4-2.3)^2(0.1) = \mathbf{0.81}$.
$\sigma_x = \sqrt{0.81} = \mathbf{0.9}$

CR7.17

Let x be the number of correct identifications. Assume that the graphologist was merely guessing, in other words that the probability of success on each trial was 0.5. Then, using Appendix Table 9 (with $n = 15$, $p = 0.5$), $P(x \geq 6) = 0.205 + 0.117 + 0.044 + 0.010 + 0.001 = 0.377$. Since this probability is not particularly small, this tells us that, if the graphologist was guessing, then it would not have been unlikely for him/her to get 6 or more correct identifications. Thus no ability to distinguish the handwriting of psychotics is indicated.

CR7.19

a $P(x > 50) = P\left(z > (50-45)/5\right) = P(z > 1) = \mathbf{0.1587}$.

b If $P(z < z^*) = 0.9$ then $z^* = 1.28$. So the required time is $\mu + (z^*)\sigma = 45 + (1.28)(5)$ $= \mathbf{51.4}$ minutes.

c If $P(z < z^*) = 0.25$ then $z^* = -0.67$. So the required time is $\mu + (z^*)\sigma = 45 + (-0.67)(5)$ $= \mathbf{41.65}$ minutes.

CR7.21

a

Survival Time (days)	Frequency	Relative Frequency
0 to <300	8	0.186
300 to <600	12	0.279
600 to <900	5	0.116
900 to <1200	5	0.116
1200 to <1500	3	0.070
1500 to <1800	3	0.070
1800 to <2100	4	0.093
2100 to <2400	1	0.023
2400 to <2700	2	0.047

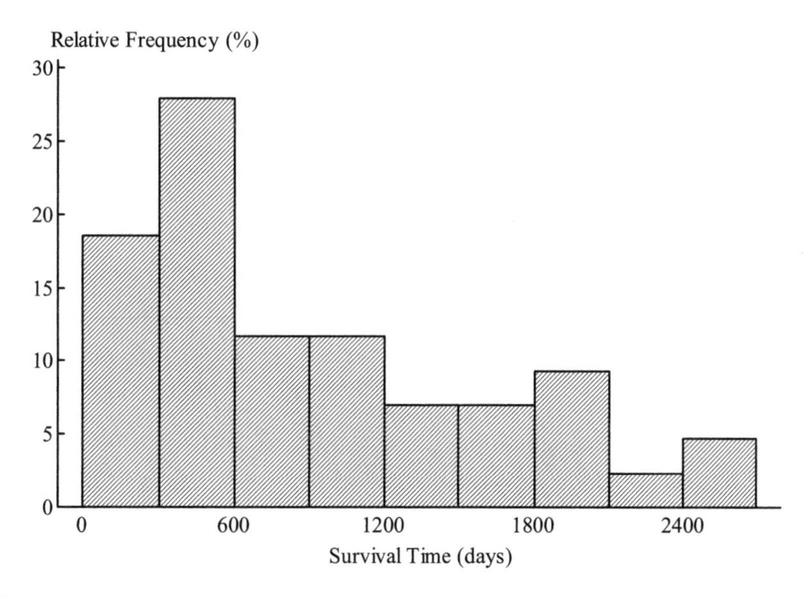

b Positive

c Yes. If a symmetrical transformation were required then a transformation would be advisable.

Chapter 8
Sampling Variability and Sampling Distributions

Note: In this chapter, numerical answers to questions involving the normal distribution were found using statistical tables. Students using calculators or computers will find that their answers differ slightly from those given.

8.1 A population characteristic is a quantity that summarizes the whole population. A statistic is a quantity calculated from the values in a sample.

8.3 **a** Population characteristic

b Statistic

c Population characteristic

d Population characteristic

e Statistic

8.5 Answers will vary.

8.7 **a**

Sample	Sample mean
1, 2	1.5
1, 3	2
1, 4	2.5
2, 1	1.5
2, 3	2.5
2, 4	3
3, 1	2
3, 2	2.5
3, 4	3.5
4, 1	2.5
4, 2	3
4, 3	3.5

The sampling distribution of the sample mean, $\bar{x}$, is shown below.

$\bar{x}$	1.5	2	2.5	3	3.5
$p(\bar{x})$	1/6	1/6	1/3	1/6	1/6

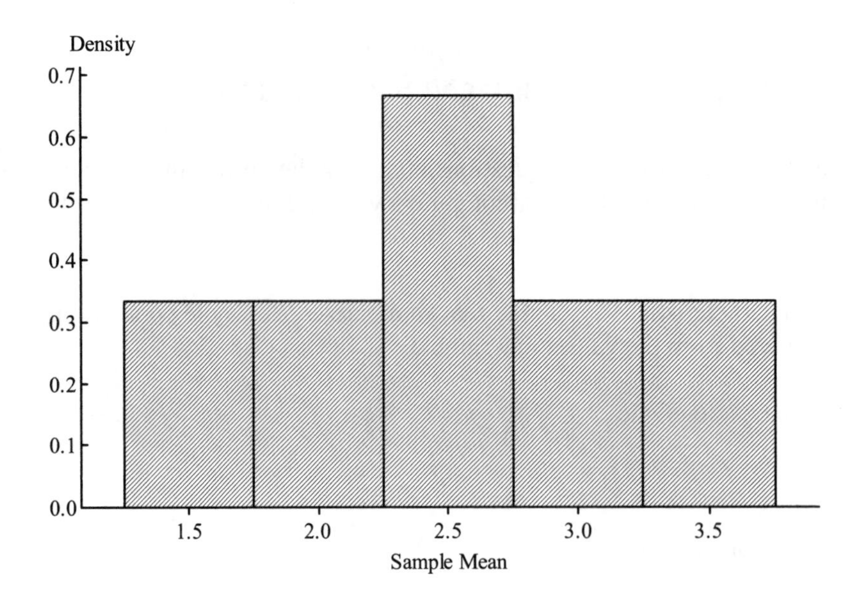

b

Sample	Sample Mean
1, 1	1
1, 2	1.5
1, 3	2
1, 4	2.5
2, 1	1.5
2, 2	2
2, 3	2.5
2, 4	3
3, 1	2
3, 2	2.5
3, 3	3
3, 4	3.5
4, 1	2.5
4, 2	3
4, 3	3.5
4, 4	4

The sampling distribution of the sample mean, $\bar{x}$, is shown below.

$\bar{x}$	1	1.5	2	2.5	3	3.5	4
$p(\bar{x})$	1/16	1/8	3/16	1/4	3/16	1/8	1/16

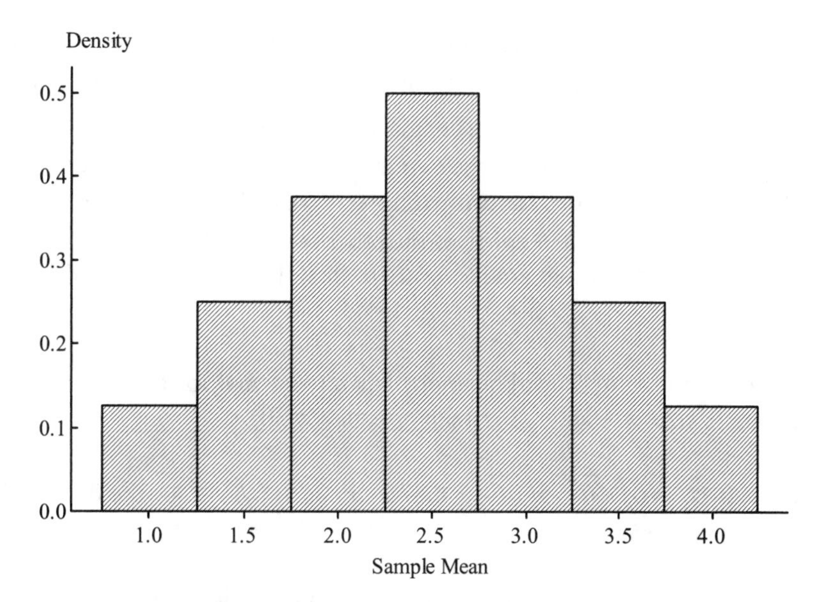

c Both distributions are symmetrical, and their means are equal (2.5). However, the "with replacement" version has a greater spread than the first distribution, with values ranging from 1 to 4 in the "with replacement" distribution, and from 1.5 to 3.5 in the "without replacement" distribution. The stepped pattern of the "with replacement" distribution more closely resembles a normal distribution than does the shape of the "without replacement" distribution.

8.9

Sample	Sample Mean	Sample Median	(Max + Min)/2
2, 3, 3*	2⅔	3	2.5
2, 3, 4	3	3	3
2, 3, 4*	3	3	3
2, 3*, 4	3	3	3
2, 3*, 4*	3	3	3
2, 4, 4*	3⅓	4	3
3, 3*, 4	3⅓	3	3.5
3, 3*, 4*	3⅓	3	3.5
3, 4, 4*	3⅔	4	3.5
3*, 4, 4*	3⅔	4	3.5

This gives probability distributions for the three statistics as shown below.

$\bar{x}$	2⅔	3	3⅓	3⅔
$p(\bar{x})$	0.1	0.4	0.3	0.2

Sample Median	3	4
p(Sample Median)	0.7	0.3

(Max + Min)/2	2.5	3	3.5
p((Max + Min)/2)	0.1	0.5	0.4

Using the sampling distributions above, the means of the three statistics are calculated to be $E(\bar{x}) = 3.2$, $E(\text{Sample median}) = 3.3$, and $E((\text{Max} + \text{Min})/2) = 3.15$. Since $\mu = 3.2$ and $E(\bar{x}) = 3.2$, we know that, on average, the sample mean will give the correct value for μ, which is not the case for either of the two other statistics. Thus, the sample mean would be the best of the three statistics for estimating μ. (Also, since the distribution of the sample mean has less variability than either of the other two distributions, the sample mean will generally produce values that are *closer* to μ than the values produced by either of the other statistics.)

8.11 The sampling distribution of $\bar{x}$ will be approximately normal for the sample sizes in Parts (c)–(f), since those sample sizes are all greater than or equal to 30.

8.13 **a** $\mu_{\bar{x}} = \mathbf{40}$, and $\sigma_{\bar{x}} = \sigma/\sqrt{n} = 5/\sqrt{64} = \mathbf{0.625}$. Since $n = 64 \geq 30$, the distribution of $\bar{x}$ will be approximately normal.

b Since $\mu - 0.5 = 40 - 0.5 = 39.5$ and $\mu + 0.5 = 40 + 0.5 = 40.5$, the required probability is $P(39.5 < \bar{x} < 40.5) = P((39.5 - 40)/0.625 < z < (40.5 - 40)/0.625) = P(-0.8 < z < 0.8)$ $= 0.7881 - 0.2119 = \mathbf{0.5762}$.

c Since $\mu - 0.7 = 40 - 0.7 = 39.3$ and $\mu + 0.7 = 40 + 0.7 = 40.7$, the probability that $\bar{x}$ will be *within* 0.7 of μ is $P(39.3 < \bar{x} < 40.7) = P((39.3 - 40)/0.625 < z < (40.7 - 40)/0.625)$ $= P(-1.12 < z < 1.12) = 0.8686 - 0.1314 = 0.7372$. Therefore, the probability that $\bar{x}$ will be more than 0.7 from μ is $1 - 0.7372 = \mathbf{0.2628}$.

8.15 **a** $\mu_{\bar{x}} = \mathbf{2}$ and $\sigma_{\bar{x}} = \sigma/\sqrt{n} = 0.8/\sqrt{9} = \mathbf{0.267}$.

b In each case $\mu_{\bar{x}} = \mathbf{2}$.
When $n = 20$, $\sigma_{\bar{x}} = \sigma/\sqrt{n} = 0.8/\sqrt{20} = \mathbf{0.179}$.
When $n = 100$, $\sigma_{\bar{x}} = \sigma/\sqrt{n} = 0.8/\sqrt{100} = \mathbf{0.08}$.
The centers of the distributions of the sample mean are all at the population mean, while the standard deviations (and therefore spreads) of these distributions are smaller for larger sample sizes. The sample size of $n = 100$ is most likely to result in a sample mean close to μ, since this is the sample size that results in the smallest standard deviation of the distribution of $\bar{x}$.

8.17 **a** Since the distribution of interpupillary distances is normal, the distribution of $\bar{x}$ is normal, also.

$$P(64 < \bar{x} < 67) = P\left(\frac{64 - 65}{\left(5/\sqrt{25}\right)} < z < \frac{67 - 65}{\left(5/\sqrt{25}\right)}\right) = P(-1 < z < 2) = 0.9772 - 0.1587 = \mathbf{0.8185}.$$

$$P(\bar{x} \geq 68) = P\left(z \geq \frac{68 - 65}{\left(5/\sqrt{25}\right)}\right) = P(z \geq 3) = \mathbf{0.0013}.$$

b Since $n = 100 \geq 30,$ the distribution of $\bar{x}$ is approximately normal.

$$P(64 < \bar{x} < 67) = P\left(\frac{64-65}{\left(5/\sqrt{100}\right)} < z < \frac{67-65}{\left(5/\sqrt{100}\right)}\right) = P(-2 < z < 4) = 1.0000 - 0.0228 = \mathbf{0.9772}.$$

$$P(\bar{x} \geq 68) = P\left(z \geq \frac{68-65}{\left(5/\sqrt{100}\right)}\right) = P(z \geq 6) = \mathbf{0.0000}.$$

8.19 Given that the true process mean is 0.5, the probability that $\bar{x}$ is *not* in the shutdown range is

$$P(0.49 < \bar{x} < 0.51) = P\left(\frac{0.49-0.5}{0.2/\sqrt{36}} < z < \frac{0.51-0.5}{0.2/\sqrt{36}}\right) = P(-3 < z < 3) = 0.9987 - 0.0013 = 0.9974.$$

So the probability that the manufacturing line will be shut down unnecessarily is $1 - 0.9974$ $= \mathbf{0.0026}$.

8.21 $$P(\text{Total} > 6000) = P(\bar{x} > 60) = P\left(z > \frac{60-50}{20/\sqrt{100}}\right) = P(z > 5) = \mathbf{0.0000}.$$

8.23 **a** $\mu_{\hat{p}} = \mathbf{0.65},\ \sigma_{\hat{p}} = \sqrt{\frac{(0.65)(0.35)}{10}} = \mathbf{0.151}.$

b $\mu_{\hat{p}} = \mathbf{0.65},\ \sigma_{\hat{p}} = \sqrt{\frac{(0.65)(0.35)}{20}} = \mathbf{0.107}.$

c $\mu_{\hat{p}} = \mathbf{0.65},\ \sigma_{\hat{p}} = \sqrt{\frac{(0.65)(0.35)}{30}} = \mathbf{0.087}.$

d $\mu_{\hat{p}} = \mathbf{0.65},\ \sigma_{\hat{p}} = \sqrt{\frac{(0.65)(0.35)}{50}} = \mathbf{0.067}.$

e $\mu_{\hat{p}} = \mathbf{0.65},\ \sigma_{\hat{p}} = \sqrt{\frac{(0.65)(0.35)}{100}} = \mathbf{0.048}.$

f $\mu_{\hat{p}} = \mathbf{0.65},\ \sigma_{\hat{p}} = \sqrt{\frac{(0.65)(0.35)}{200}} = \mathbf{0.034}.$

8.25 **a** $\mu_{\hat{p}} = \mathbf{0.07},\ \sigma_{\hat{p}} = \sqrt{\frac{(0.07)(0.93)}{100}} = \mathbf{0.026}.$

b No, since $np = 100(0.07) = 7,$ which is not greater than or equal to 10.

c The mean is unchanged since the mean of the distribution of the sample proportion is always equal to the population proportion, but the standard deviation changes to
$\sigma_{\hat{p}} = \sqrt{\frac{(0.07)(0.93)}{200}} = \mathbf{0.018}.$

d Yes, since $np = 200(0.07) = 14$ and $n(1-p) = 200(0.93) = 186$, which are both greater than or equal to 10.

e $P(\hat{p} > 0.1) = P\left(z > \frac{0.1 - 0.07}{\sqrt{(0.07)(0.93)/200}}\right) = P(z > 1.66) = \mathbf{0.0485}.$

8.27 **a** $\mu_{\hat{p}} = \frac{1}{200} = \mathbf{0.005},\ \sigma_{\hat{p}} = \sqrt{\frac{(0.005)(0.995)}{100}} = \mathbf{0.007}.$

b No, since $np = 100(0.005) = 0.5$, which is not greater than or equal to 10.

c We need both np and $n(1-p)$ to be greater than or equal to 10, and since $p < q$ it will be sufficient to ensure that $np \geq 10$. So we need $n(0.005) \geq 10$, that is $n \geq 10/0.005 = \mathbf{2000}$.

8.29 **a** If $p = 0.5$, $\mu_{\hat{p}} = \mathbf{0.5}$, $\sigma_{\hat{p}} = \sqrt{\frac{(0.5)(0.5)}{225}} = \mathbf{0.0333}$. Also $np = 225(0.5) = 112.5 \geq 10$ and $n(1-p) = 225(0.5) = 112.5 \geq 10$, and so $\hat{p}$ has an approximately normal distribution.
If $p = 0.6$, $\mu_{\hat{p}} = \mathbf{0.6}$, $\sigma_{\hat{p}} = \sqrt{\frac{(0.6)(0.4)}{225}} = \mathbf{0.0327}$. Also $np = 225(0.6) = 135 \geq 10$ and $n(1-p) = 225(0.4) = 90 \geq 10$, and so $\hat{p}$ has an approximately normal distribution.

b If $p = 0.5$, $P(\hat{p} \geq 0.6) = P\left(z \geq \frac{0.6 - 0.5}{\sqrt{(0.5)(0.5)/225}}\right) = P(z \geq 3) = \mathbf{0.0013}.$
If $p = 0.6$, $P(\hat{p} \geq 0.6) = P\left(z \geq \frac{0.6 - 0.6}{\sqrt{(0.6)(0.4)/225}}\right) = P(z \geq 0) = \mathbf{0.5}.$

c For a larger sample size, the value of $\hat{p}$ is likely to be closer to p. So, for $n = 400$, when $p = 0.5$, $P(\hat{p} \geq 0.6)$ will be **smaller**. When $p = 0.6$, $P(\hat{p} \geq 0.6)$ will still be 0, that is, it will remain **the same**.

8.31 **a** $P(\text{Returned}) = P(\hat{p} > 0.02) = P\left(z > \frac{0.02 - 0.05}{\sqrt{(0.05)(0.95)/200}}\right) = P(z > -1.95) = \mathbf{0.9744}.$

b $P(\text{Returned}) = P(\hat{p} > 0.02) = P\left(z > \frac{0.02 - 0.1}{\sqrt{(0.1)(0.9)/200}}\right) = P(z > -3.77) = 0.9999.$ So the probability that the shipment is not returned is $1 - 0.9999 = \mathbf{0.0001}$.

8.33 **a** Since $n = 100 \geq 30$, $\bar{x}$ is approximately normally distributed. Its mean is 50 and its standard deviation is $\sqrt{1}/\sqrt{100} = 0.1$.

b $P(49.75 < \bar{x} < 50.25) = P\left(\frac{49.75-50}{0.1} < z < \frac{50.25-50}{0.1}\right) = P(-2.5 < z < 2.5) = 0.9938 - 0.0062$
$= \mathbf{0.9876}$.

c Since $\mu_{\bar{x}} = 50$, $P(\bar{x} < 50) = \mathbf{0.5}$.

8.35 **a** Let the index of the specimen be x. $P(850 < x < 1300) = P\left(\frac{850-1000}{150} < z < \frac{1300-1000}{150}\right)$
$= P(-1 < z < 2) = 0.9772 - 0.1587 = \mathbf{0.8185}$.

b $P(950 < \bar{x} < 1100) = P\left(\frac{950-1000}{150/\sqrt{10}} < z < \frac{1100-1000}{150/\sqrt{10}}\right) = P(-1.05 < z < 2.11)$
$= 0.9826 - 0.1469 = \mathbf{0.8357}$.
$P(850 < \bar{x} < 1300) = P\left(\frac{850-1000}{150/\sqrt{10}} < z < \frac{1300-1000}{150/\sqrt{10}}\right) = P(-3.16 < z < 6.32)$
$= 1.0000 - 0.0008 = \mathbf{0.9992}$.

8.37 $P(\text{Total} > 5300) = P(\bar{x} > 5300/50) = P(\bar{x} > 106) = P\left(z > \frac{106-100}{30/\sqrt{50}}\right) = P(z > 1.41) = \mathbf{0.0793}$.

Chapter 9
Estimation Using a Single Sample

Note: In this chapter, numerical answers to questions involving the normal and t distributions were found using statistical tables. Students using calculators or computers will find that their answers differ slightly from those given.

9.1 Statistics II and III are preferable to Statistic I since they are unbiased (their means are equal to the value of the population characteristic). However, Statistic II is preferable to Statistic III since its standard deviation is smaller. So **Statistic II** should be recommended.

9.3 $\hat{p} = 1720/6212 = \mathbf{0.277}$.

9.5 The value of p is estimated using $\hat{p}$, and the value of $\hat{p}$ is $14/20 = \mathbf{0.7}$.

9.7 **a** The value of μ is estimated using $\bar{x} = (410 + \cdots + 530)/7 = \mathbf{421.429}$.

b The value of σ^2 is estimated using $s^2 = \mathbf{10414.286}$.

c The value of σ is estimated using $s = \mathbf{102.050}$. No, s is not an unbiased statistic for estimating σ.

9.9 **a** The value of μ_J is estimated using $\bar{x} = (103 + \cdots + 99)/10 = \mathbf{120.6}$ therms.

b The value of τ is estimated to be 10000(120.6) = **1,206,000** therms.

c The value of p is estimated using $\hat{p} = 8/10 = \mathbf{0.8}$.

d The population median is estimated using the sample median, which is **120** therms.

9.11 **a** **1.96**

b **1.645**

c **2.58**

d **1.28**

e **1.44**

9.13 **a** The larger the confidence level the wider the interval.

b The larger the sample size the narrower the interval.

c Values of $\hat{p}$ further from 0.5 give smaller values of $\hat{p}(1-\hat{p})$. Therefore, the further the value of $\hat{p}$ from 0.5, the narrower the interval.

9.15 Check of Conditions

1. Since $n\hat{p} = 1100(990/1100) = 990 \geq 10$ and $n(1-\hat{p}) = 1100(110/1100) = 110 \geq 10$, the sample size is large enough.
2. The sample size of n = 1100 is much smaller than 10% of the population size (the number of drivers).
3. We are told to assume that the sample is representative of the population of drivers. Having made this assumption it is reasonable to regard the sample as a random sample from the population.

Calculation

The 99% confidence interval for p is

$$\hat{p} \pm 2.58\sqrt{\frac{\hat{p}(1-\hat{p})}{n}} = \frac{990}{1100} \pm 2.58\sqrt{\frac{(990/1100)(110/1100)}{1100}} = (\mathbf{0.877, 0.923}).$$

Interpretation

We are 99% confident that the proportion of all drivers who have engaged in careless or aggressive driving in the last six months is between 0.877 and 0.923.

9.17 Let p be the proportion of all coastal residents who *would* evacuate.

Check of Conditions

1. Since $n\hat{p} = 5046(0.69) = 3482 \geq 10$ and $n(1-\hat{p}) = 5046(0.31) = 1564 \geq 10$, the sample size is large enough.
2. The sample size of n = 5046 is much smaller than 10% of the population size (the number of people who live within 20 miles of the coast in high hurricane risk counties of these eight southern states).
3. The sample was selected in a way designed to produce a representative sample. So, it is reasonable to regard the sample as a random sample from the population.

Calculation

The 98% confidence interval for p is

$$\hat{p} \pm 2.33\sqrt{\frac{\hat{p}(1-\hat{p})}{n}} = 0.69 \pm 2.33\sqrt{\frac{(0.69)(0.31)}{5046}} = (\mathbf{0.675, 0.705}).$$

Interpretation of the Confidence Interval

We are 98% confident that the proportion of all coastal residents who would evacuate is between 0.675 and 0.705.

Interpretation of the Confidence Level

If we were to take a large number of random samples of size 5046, 98% of the resulting confidence intervals would contain the true proportion of all coastal residents who would evacuate.

9.19 **a** Check of Conditions

1. Since $n\hat{p} = 2002(1321/2002) = 1321 \geq 10$ and $n(1-\hat{p}) = 2002(681/2002) = 681 \geq 10$, the sample size is large enough.
2. The sample size of n = 2002 is much smaller than 10% of the population size (the number of Americans age 8 to 18).
3. The sample was selected in a way designed to produce a representative sample. So, it is reasonable to regard the sample as a random sample from the population.

Calculation

The 90% confidence interval for p is

$$\hat{p} \pm 1.645\sqrt{\frac{\hat{p}(1-\hat{p})}{n}} = \frac{1321}{2002} \pm 1.645\sqrt{\frac{(1321/2002)(681/2002)}{2002}} = (\mathbf{0.642, 0.677}).$$

Interpretation

We are 90% confident that the proportion of all Americans age 8 to 18 who own a cell phone is between 0.642 and 0.677.

b Check of Conditions

1. Since $n\hat{p} = 2002(1522/2002) = 1522 \geq 10$ and $n(1-\hat{p}) = 2002(480/2002) = 480 \geq 10$, the sample size is large enough.
2. The sample size of $n = 2002$ is much smaller than 10% of the population size (the number of Americans age 8 to 18).
3. The sample was selected in a way designed to produce a representative sample. So it is reasonable to regard the sample as a random sample from the population.

Calculation

The 90% confidence interval for p is

$$\hat{p} \pm 1.645\sqrt{\frac{\hat{p}(1-\hat{p})}{n}} = \frac{1522}{2002} \pm 1.645\sqrt{\frac{(1522/2002)(480/2002)}{2002}} = (\mathbf{0.745, 0.776}).$$

Interpretation

We are 90% confident that the proportion of all Americans age 8 to 18 who own an MP3 player is between 0.745 and 0.776.

c The interval in Part (b) is narrower than the interval in Part (a) because the sample proportion in Part (b) is further from 0.5, thus reducing the value of the estimated standard deviation of the sample proportion (given by the expression $\sqrt{\hat{p}(1-\hat{p})/n}$).

9.21 a Check of Conditions

1. Since $n\hat{p} = 500(350/500) = 350 \geq 10$ and $n(1-\hat{p}) = 500(150/500) = 150 \geq 10$, the sample size is large enough.
2. The sample size of $n = 500$ is much smaller than 10% of the population size (the number of potential jurors).
3. We are told to assume that the sample is representative of the population of potential jurors. Having made this assumption it is reasonable to regard the sample as a random sample from the population.

Calculation

The 95% confidence interval for p is

$$\hat{p} \pm 1.96\sqrt{\frac{\hat{p}(1-\hat{p})}{n}} = \frac{350}{500} \pm 1.96\sqrt{\frac{(350/500)(150/500)}{500}} = (\mathbf{0.660, 0.740}).$$

Interpretation

We are 95% confident that the proportion of all potential jurors who regularly watch at least one crime-scene investigation series is between 0.660 and 0.740.

b Wider

9.23 a Check of Conditions

1. Since $n\hat{p} = 526(137/526) = 137 \geq 10$ and $n(1-\hat{p}) = 526(389/526) = 389 \geq 10$, the sample size is large enough.
2. The sample size of $n = 526$ is much smaller than 10% of the population size (the number of U.S. businesses).
3. We must assume that the sample is a random sample of U.S. businesses.

Calculation

The 95% confidence interval for p is

$$\hat{p} \pm 1.96\sqrt{\frac{\hat{p}(1-\hat{p})}{n}} = \frac{137}{526} \pm 1.96\sqrt{\frac{(137/526)(389/526)}{526}} = (\mathbf{0.223}, \mathbf{0.298}).$$

Interpretation

We are 95% confident that the proportion of all U.S. businesses that have fired workers for misuse of the Internet is between 0.223 and 0.298.

b The sample proportion of businesses that had fired workers for misuse of email is further from 0.5 than the sample proportion of businesses that had fired workers for misuse of the Internet, making the value of $\sqrt{\hat{p}(1-\hat{p})/n}$ smaller. This makes the confidence interval narrower. Additionally, the critical value of z for a 90% confidence interval is smaller than the critical value of z for a 95% confidence interval, also making the second confidence interval narrower.

9.25 Check of Conditions

1. Since $n\hat{p} = 1002(0.82) = 822 \geq 10$ and $n(1-\hat{p}) = 1002(0.18) = 180 \geq 10$, the sample size is large enough.
2. The sample size of n = 1002 is much smaller than 10% of the population size (the number of adults in the country).
3. We are told that the sample was a random sample from the population.

Calculation

When the sample proportion of 0.82 is used as an estimate of the population proportion, the 95% error bound on this estimate is

$$1.96\sqrt{\frac{\hat{p}(1-\hat{p})}{n}} = 1.96\sqrt{\frac{(0.82)(0.18)}{1002}} = 0.024.$$

Interpretation

We are 95% confident that proportion of all adults who believe that the shows are either "totally made up" or "mostly distorted" is within 2.4% of the sample proportion of 82%.

9.27 We are 95% confident that proportion of all adult drivers who would say that they often or sometimes talk on a cell phone while driving is within $1.96\sqrt{\hat{p}(1-\hat{p})/n}$ $= 1.96\sqrt{(0.36)(0.64)/1004} = 0.030$, that is, 3.0 percentage points, of the sample proportion of 36%. The reported bound on error is slightly inaccurate, in that it is wrong by one tenth of a percentage point.

9.29 **a** Check of Conditions

1. Since $n\hat{p} = 89(18/89) = 18 \geq 10$ and $n(1-\hat{p}) = 89(71/89) = 71 \geq 10$, the sample size is large enough.
2. The sample size of n = 89 is much smaller than 10% of the population size (the number of people under 50 years old who use this type of defibrillator).
3. We are told to assume that the sample is representative of patients under 50 years old who receive this type of defibrillator. Having made this assumption it is reasonable to regard the sample as a random sample from the population.

Calculation

The 95% confidence interval for p is

$$\hat{p} \pm 1.96\sqrt{\frac{\hat{p}(1-\hat{p})}{n}} = \frac{18}{89} \pm 1.96\sqrt{\frac{(18/89)(71/89)}{89}} = (\mathbf{0.119, 0.286}).$$

Interpretation

We are 95% confident that the proportion of all patients under 50 years old who experience a failure within the first two years after receiving this type of defibrillator is between 0.119 and 0.286.

b Check of Conditions

1. Since $n\hat{p} = 362(13/362) = 13 \geq 10$ and $n(1-\hat{p}) = 362(349/362) = 349 \geq 10$, the sample size is large enough.
2. The sample size of $n = 362$ is much smaller than 10% of the population size (the number of people age 50 or older who use this type of defibrillator).
3. We are told to assume that the sample is representative of patients age 50 or older who receive this type of defibrillator. Having made this assumption it is reasonable to regard the sample as a random sample from the population.

Calculation

The 99% confidence interval for p is

$$\hat{p} \pm 2.58\sqrt{\frac{\hat{p}(1-\hat{p})}{n}} = \frac{13}{362} \pm 2.58\sqrt{\frac{(13/362)(349/362)}{362}} = (\mathbf{0.011, 0.061}).$$

Interpretation

We are 99% confident that the proportion of all patients age 50 or older who experience a failure within the first two years after receiving this type of defibrillator is between 0.011 and 0.061.

c Using the estimate of p from the study, $18/89$, the required sample size is given by

$$n = p(1-p)\left(\frac{1.96}{B}\right)^2 = \left(\frac{18}{89}\right)\left(\frac{71}{89}\right)\left(\frac{1.96}{0.03}\right)^2 = 688.685.$$

So a sample of size at least **689** is required.

9.31 $n = p(1-p)\left(\frac{1.96}{B}\right)^2 = 0.25\left(\frac{1.96}{0.02}\right)^2 = 2401$. A sample size of **2401** is required.

9.33 $n = p(1-p)\left(\frac{1.96}{B}\right)^2 = 0.25\left(\frac{1.96}{0.05}\right)^2 = 384.16$. A sample size of **385** is required.

9.35 **a** **2.12**

b **1.80**

c **2.81**

d **1.71**

e **1.78**

f **2.26**

9.37 The width of the first interval is $52.7-51.3=1.4$. The width of the second interval is $50.6-49.4=1.2$. Since the confidence interval is given by $\bar{x}\pm(t\text{ critical value})\left(s/\sqrt{n}\right)$, the *width* of the confidence interval is given by $2\cdot(t\text{ critical value})\left(s/\sqrt{n}\right)$. Therefore, for samples of equal standard deviations, the larger the sample size the narrower the interval. Thus it is the **second** interval that is based on the larger sample size.

9.39 **a** Conditions

1. Since $n=411\geq 30$, the sample size is large enough.
2. We are told to assume that the sample is representative of students taking introductory psychology at this university. Having made this assumption it is reasonable to regard the sample as a random sample from the population.

Calculation

The 95% confidence interval for μ is

$$\bar{x}\pm(t\text{ critical value})\cdot\frac{s}{\sqrt{n}}=7.74\pm1.96\cdot\frac{3.40}{\sqrt{411}}=\mathbf{(7.411, 8.069)}.$$

Interpretation

We are 95% confident that the mean time spent studying for this exam for all students taking introductory psychology at this university is between 7.411 and 8.069 hours.

b Conditions

1. Since $n=411\geq 30$, the sample size is large enough.
2. We are told to assume that the sample is representative of students taking introductory psychology at this university. Having made this assumption it is reasonable to regard the sample as a random sample from the population.

Calculation

The 90% confidence interval for μ is

$$\bar{x}\pm(t\text{ critical value})\cdot\frac{s}{\sqrt{n}}=43.18\pm1.645\cdot\frac{21.46}{\sqrt{411}}=\mathbf{(41.439, 44.921)}.$$

Interpretation

We are 90% confident that the mean percent of study time that occurs in the 24 hours prior to the exam for all students taking introductory psychology at this university is between 41.439 and 44.921.

9.41 **a** The fact that the mean is much greater than the median suggests that the distribution of times spent volunteering in the sample was positively skewed.

b With the sample mean being much greater than the sample median, and with the sample being regarded as representative of the population, it seems very likely that the population is strongly positively skewed, and therefore not normally distributed.

c Since $n=1086\geq 30$, the sample size is large enough for us to use the t confidence interval, even though the population distribution is not approximately normal.

d In addition to observing that the sample is large, we need to point out that the sample was selected in a way that makes it reasonable to regard it as representative of the population, and therefore that it is reasonable to regard the sample as random. This justifies use of the t confidence interval. The 98% confidence interval for μ is then

$$\bar{x} \pm (t \text{ critical value}) \cdot \frac{s}{\sqrt{n}} = 5.6 \pm 2.33 \cdot \frac{5.2}{\sqrt{1086}} = \mathbf{(5.232, 5.968)}.$$

We are 98% confident that the mean time spent volunteering for the population of parents of school age children is between 5.232 and 5.968 hours.

9.43 **a** The 90% confidence interval would be **narrower**. In order to be only 90% confident that the interval captures the true population mean, the interval does not have to be as wide as it would in order to be 95% confident of capturing the true population mean.

b The statement is not correct. The population mean, μ, is a constant, and therefore we cannot talk about the probability that it falls within a certain interval.

c The statement is not correct. We can say that *on average* 95 out of every 100 samples will result in confidence intervals that will contain μ, but we cannot say that in 100 such samples, *exactly* 95 will result in confidence intervals that contain μ.

9.45 **a** For samples of equal sizes, those with greater variability will result in wider confidence intervals. The 12 to 23 month and 24 to 35 month samples resulted in confidence intervals of width 0.4, while the less than 12 month sample resulted in a confidence interval of width 0.2. So the 12 to 23 month and 24 to 35 month samples are the ones with the greater variability.

b For samples of equal variability, those with greater sample sizes will result in narrower confidence intervals. Thus the less than 12 month sample is the one with the greater sample size.

c Since the new interval is wider than the interval given in the question, the new interval must be for a higher confidence level. (By obtaining a wider interval, we have a greater confidence that the interval captures the true population mean.) Thus the new interval must have a **99%** confidence level.

9.47 **a** Conditions

1. Since $n = 100 \geq 30$, the sample size is large enough.
2. We are told to assume that the sample was a random sample of passengers.

Calculation

The t critical value for 99 degrees of freedom (for a 95% confidence level) is between 1.98 and 2.00. We will use an estimate of 1.99. Thus, the 95% confidence interval for μ is

$$\bar{x} \pm (t \text{ critical value}) \cdot \frac{s}{\sqrt{n}} = 183 \pm 1.99 \cdot \frac{20}{\sqrt{100}} = \mathbf{(179.02, 186.98)}.$$

Interpretation

We are 95% confident that the mean summer weight is between 179.02 and 186.98 lb.

b Conditions

1. Since $n = 100 \geq 30$, the sample size is large enough.
2. We are told to assume that the sample was a random sample.

Calculation

The 95% confidence interval for μ is

$$\bar{x} \pm (t \text{ critical value}) \cdot \frac{s}{\sqrt{n}} = 190 \pm 1.99 \cdot \frac{23}{\sqrt{100}} = \mathbf{(185.423, 194.577)}.$$

Interpretation
We are 95% confident that the mean winter weight is between 185.423 and 194.577 lb.

c Based on the Frontier Airlines data, neither recommendation is likely to be an accurate estimate of the mean passenger weight, since 190 is not contained in the confidence interval for the mean summer weight and 95 is not contained in the confidence interval for the mean winter weight.

9.49 A boxplot of the sample values is shown below.

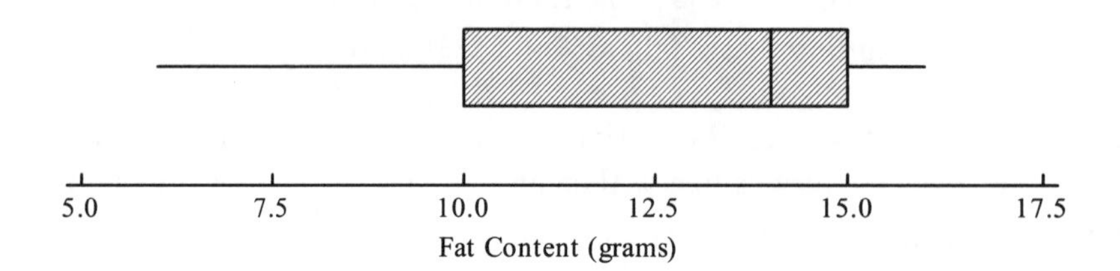

The boxplot shows that the distribution of the sample values is negatively skewed, and this leads us to suspect that the population is not approximately normally distributed. Therefore, since the sample is small, it is not appropriate to use the *t* confidence interval method of this section.

9.51 A reasonable estimate of σ is given by $(\text{sample range})/4 = (700-50)/4 = 162.5.$ Thus

$$n = \left(\frac{1.96\sigma}{B}\right)^2 = \left(\frac{1.96 \cdot 162.5}{10}\right)^2 = 1014.4225.$$

So we need a sample size of **1015**.

9.53 First, we need to know that the information is based on a random sample of middle-income consumers aged 65 and older. Second, it would be useful if some sort of margin of error were given for the estimated mean of \$10,235.

9.55 **a** The paper states that Queens flew for an *average of* 24.2 ± 9.21 minutes on their mating flights, and so this interval is a confidence interval for a population mean.

b Conditions

1. Since $n = 30 \geq 30,$ the sample size is large enough.
2. We are told to assume that the 30 queen honeybees are representative of the population of queen honeybees. It is then reasonable to treat the sample as a random sample from the population.

Calculation
The 95% confidence interval for μ is

$$\bar{x} \pm (t \text{ critical value}) \cdot \frac{s}{\sqrt{n}} = 4.6 \pm 2.05 \cdot \frac{3.47}{\sqrt{30}} = \mathbf{(3.301,\ 5.899)}.$$

Interpretation
We are 95% confident that the mean number of partners is between 3.301 and 5.899.

9.57 Check of Conditions

1. Since $n\hat{p} = 52(18/52) = 18 \geq 10$ and $n(1-\hat{p}) = 52(34/52) = 34 \geq 10$, the sample size is large enough.
2. The sample size of $n = 52$ is much smaller than 10% of the population size (the number of young adults with pierced tongues).
3. The assumption we have made is that the sample of 52 was a random sample from the population of young adults with pierced tongues.

Calculation

The 95% confidence interval for p is

$$\hat{p} \pm 1.96\sqrt{\frac{\hat{p}(1-\hat{p})}{n}} = \frac{18}{52} \pm 1.96\sqrt{\frac{(18/52)(34/52)}{52}} = (\mathbf{0.217}, \mathbf{0.475}).$$

Interpretation

We are 95% confident that the proportion of all young people with pierced tongues who have receding gums is between 0.217 and 0.475.

9.59 The standard error for the mean cost for Native Americans is much larger than that for Hispanics since the sample size was much smaller for Native Americans.

9.61 Check of Conditions

1. Since $n\hat{p} = 150(0.65) = 97.5 \geq 10$ and $n(1-\hat{p}) = 150(0.35) = 52.5 \geq 10$, the sample size is large enough.
2. The sample size of $n = 150$ is much smaller than 10% of the population size (the number of Utah residents).
3. We are told to assume that the sample was a random sample from the population.

Calculation

The 90% confidence interval for p is

$$\hat{p} \pm 1.645\sqrt{\frac{\hat{p}(1-\hat{p})}{n}} = 0.65 \pm 1.645\sqrt{\frac{(0.65)(0.35)}{150}} = (\mathbf{0.586}, \mathbf{0.714}).$$

Interpretation

We are 90% confident that the proportion of all Utah residents who favor fluoridation is between 0.586 and 0.714.

Yes. Since the whole of this interval is above 0.5, the interval is consistent with the statement that fluoridation is favored by a clear majority of Utah residents.

9.63 Check of Conditions

1. Since $n\hat{p} = 750(125/750) = 125 \geq 10$ and $n(1-\hat{p}) = 750(625/750) = 625 \geq 10$, the sample size is large enough.
2. The sample size of $n = 750$ is much smaller than 10% of the population size (the number of full-time workers).
3. We are told to assume that the sample is a random sample from the population of full-time workers.

Calculation

The 90% confidence interval for p is

$$\hat{p} \pm 1.645\sqrt{\frac{\hat{p}(1-\hat{p})}{n}} = \frac{125}{750} \pm 1.645\sqrt{\frac{(125/750)(625/750)}{750}} = (\mathbf{0.144}, \mathbf{0.189}).$$

Interpretation
We are 90% confident that the proportion of all full-time workers who have been so angered in the last year that they wanted to hit a colleague is between 0.144 and 0.189.

9.65 $n = p(1-p)\left(\frac{1.96}{B}\right)^2 = (0.5)(0.5)\left(\frac{1.96}{0.1}\right)^2 = 96.04$. A sample size of **97** is required.

9.67 $n = \left(\frac{1.96\sigma}{B}\right)^2 = \left(\frac{1.96 \cdot 0.8}{0.1}\right)^2 = 245.862$. A sample size of **246** is required.

9.69 The 99% upper confidence bound for the mean wait time for bypass surgery is $19 + 2.33\left(10/\sqrt{539}\right) = \mathbf{20.004}$ days.

9.71 The 95% confidence interval for the population standard deviation of wait time (in days) for angiography is

$$9 \pm 1.96\left(\frac{9}{\sqrt{2(847)}}\right) = \mathbf{(8.571, 9.429)}.$$

9.73 Conditions
1. We have to assume that the distribution of the time taken to eat a frog over all Indian false vampire bats is normally distributed.
2. We have to assume, also, that the sample of 12 bats is a *random* sample from the population of Indian false vampire bats.

Calculation
The 90% confidence interval for μ is

$$\bar{x} \pm (t \text{ critical value}) \cdot \frac{s}{\sqrt{n}} = 21.9 \pm 1.80 \cdot \frac{7.7}{\sqrt{12}} = \mathbf{(17.899, 25.901)}.$$

Interpretation
We are 90% confident that the mean suppertime for a vampire bat whose meal consists of a frog is between 17.899 and 25.901 minutes.

Chapter 10
Hypothesis Testing Using a Single Sample

Note: In this chapter, numerical answers to questions involving the normal and *t* distributions were found using values from a calculator. Students using statistical tables will find that their answers differ slightly from those given.

10.1 Legitimate hypotheses concern *population* characteristics; $\bar{x}$ is a sample statistic.

10.3 Because so much is at stake at a nuclear power plant, the inspection team needs to obtain convincing evidence that everything is in order. To put this another way, the team needs not only to obtain a sample mean greater than 100 but, beyond that, to be sure that sample mean is sufficiently far above 100 to provide convincing evidence that the true mean weld strength is greater than 100. Hence an alternative hypothesis of $H_a: \mu > 100$ will be used.

10.5 We are clearly talking here about a situation where, in a sample of children who had received the MMR vaccine, a higher incidence of autism was observed than the incidence of autism in children in general. The process of the hypothesis test is then to assume that the incidence of autism is the same amongst the population of children who have had the MMR vaccine as it is amongst children in general, and then to find out whether, on that basis, a result such as the one obtained in the sample would be very unusual, or not particularly unusual. If such a result would be very unusual, then the sample result is providing convincing evidence of a higher incidence of autism amongst the population of children who have received the MMR vaccine than in children in general. If the sample result would not be particularly unusual, then it would *not* provide convincing evidence of this. However, since the incidence of autism amongst children in the sample was observed to be higher than it is known to be in children in general, there's no way that this result can provide evidence that MMR does *not* cause autism.

10.7 We assume that the program director will continue with the station's current programming unless there is convincing evidence that more that half of the potential viewers prefer a return to the regular programming. Thus, letting *p* be the proportion of all potential viewers who would prefer a return to the regular programming, the program director should test H_0: $p = 0.5$ versus H_a: $p > 0.5$.

10.9 Let *p* be the proportion of all constituents who favor spending money for the new sewer system. She should test H_0: $p = 0.5$ versus H_a: $p > 0.5$.

10.11 Let μ be the population mean amperage at which the fuses burn out. Action will need to be taken if the data provide convincing evidence that either $\mu < 40$ or $\mu > 40$. Thus the manufacturer should test H_0: $\mu = 40$ versus H_a: $\mu \neq 40$.

10.13 **a** This is a **Type I** error. Its probability is $3/33 = \mathbf{0.091}$.

b A Type II error would be coming to the conclusion that the woman has cancer in the other breast when in fact she does not have cancer in the other breast. The probability that this happens is $91/936 = \mathbf{0.097}$.

10.15 **a** A Type I error would be coming to the conclusion that the man is not the father when in fact he is. A Type II error would be not coming to the conclusion that the man is not the father when in fact he is not the father.

b $\alpha = 0.001$, $\beta = 0$.

c A "false positive" is coming to the conclusion that the man is the father when in fact he is not the father. This is a Type II error, and its probability is $\beta = 0.008$.

10.17 **a** A Type I error is obtaining convincing evidence that more than 1% of a shipment is defective when in fact (at least) 1% of the shipment is defective. A Type II error is not obtaining convincing evidence that more than 1% of a shipment is defective when in fact more than 1% of the shipment is defective.

b The consequence of a Type I error would be that the calculator manufacturer returns a shipment when in fact it was acceptable. This will do minimal harm to the calculator manufacturer's business. However, the consequence of a Type II error would be that the calculator manufacturer would go ahead and use in the calculators circuits that are defective. This will then lead to faulty calculators and would therefore be harmful to the manufacturer's business. A **Type II** error would be the more serious for the calculator manufacturer.

c At least in the short term, a Type II error would not be harmful to the supplier's business; payment would be received for a shipment that was in fact faulty. However, if a Type I error were to occur, the supplier would receive back, and not be paid for, a shipment of circuits that was in fact acceptable. A **Type I** error would be the more serious for the supplier.

10.19 **a** Before filing charges of false advertising against the company, the consumer advocacy group would require convincing evidence that more than 10% of the flares are defective.

b A Type I error is coming to the conclusion that more than 10% of the flares are defective when in fact 10% (or fewer) of the flares are defective. This would result in the expensive and time-consuming process of filing charges of false advertising against the company when in fact the company is not at fault. A Type II error is not coming to the conclusion that more than 10% of the flares are defective when in fact more than 10% of the flares are defective. As a result the consumer advocacy group would not file charges when in fact the company was at fault.

10.21 **a** The researchers failed to reject H_0.

b If the researchers were incorrect in their conclusion, then they would be failing to reject H_0 when H_0 was in fact true. This is a **Type II** error.

c Yes. The study did not provide convincing evidence that there is a higher cancer death rate for people who live close to nuclear facilities. However, this does not mean that there was *no* such effect, and this would be the case for any study with the same outcome.

10.23 **a** A *P*-value of 0.0003 means that it is very unlikely (probability = 0.0003), assuming that H_0 is true, that you would get a sample result at least as inconsistent with H_0 as the one obtained in the study. Thus H_0 is rejected.

b A P-value of 0.350 means that it is not particularly unlikely (probability = 0.350), assuming that H_0 is true, that you would get a sample result at least as inconsistent with H_0 as the one obtained in the study. Thus there is no reason to reject H_0.

10.25 **a** H_0 is not rejected.

b H_0 is not rejected.

c H_0 is not rejected.

d H_0 is rejected.

e H_0 is not rejected.

f H_0 is not rejected.

10.27 **a** The large-sample z test is **not** appropriate since $np = 25(0.2) = 5 < 10$.

b The large-sample z test **is** appropriate since $np = 210(0.6) = 126 \geq 10$ and $n(1-p) = 210(0.4) = 84 \geq 10$.

c The large-sample z test **is** appropriate since $np = 100(0.9) = 90 \geq 10$ and $n(1-p) = 100(0.1) = 10 \geq 10$.

d The large-sample z test is **not** appropriate since $np = 75(0.05) = 3.75 < 10$.

10.29 **a**

1. p = proportion of all women who work full time, age 22 to 35, who would be willing to give up some personal time in order to make more money.
2. H_0: $p = 0.5$
3. H_a: $p > 0.5$
4. $\alpha = 0.01$
5. $z = \dfrac{\hat{p} - p}{\sqrt{\dfrac{p(1-p)}{n}}} = \dfrac{\hat{p} - 0.5}{\sqrt{\dfrac{(0.5)(0.5)}{1000}}}$
6. The sample was selected in a way that was designed to produce a sample that was representative of women in the targeted group, so it is reasonable to treat the sample as a random sample from the population. The sample size is much smaller than the population size (the number of women age 22 to 35 who work full time). Furthermore, $np = 1000(0.5) = 500 \geq 10$ and $n(1-p) = 1000(0.5) = 500 \geq 10$, so the sample is large enough. Therefore the large sample test is appropriate.
7. $z = \dfrac{540/1000 - 0.5}{\sqrt{\dfrac{(0.5)(0.5)}{1000}}} = 2.52982$
8. $P\text{-value} = P(Z > 2.52982) = 0.00571$
9. Since $P\text{-value} = 0.00571 < 0.01$ we reject H_0. We have convincing evidence that a majority of women age 22 to 35 who work full time would be willing to give up some personal time for more money.

b No. The survey only covered women age 22 to 35.

10.31
1. p = proportion of all adult Americans who would prefer to live in a hot climate rather than a cold climate
2. H_0: $p = 0.5$
3. H_a: $p > 0.5$
4. $\alpha = 0.01$
5. $z = \dfrac{\hat{p} - p}{\sqrt{\dfrac{p(1-p)}{n}}} = \dfrac{\hat{p} - 0.5}{\sqrt{\dfrac{(0.5)(0.5)}{2260}}}$
6. The sample was nationally representative, so it is reasonable to treat the sample as a random sample from the population. The sample size is much smaller than the population size (the number of adult Americans). Furthermore, $np = 2260(0.5) = 1130 \geq 10$ and $n(1-p) = 2260(0.5) = 1130 \geq 10$, so the sample is large enough. Therefore the large sample test is appropriate.
7. $z = \dfrac{1288/2260 - 0.5}{\sqrt{\dfrac{(0.5)(0.5)}{2260}}} = 6.64711$
8. $P\text{-value} = P(Z > 6.64711) \approx 0$
9. Since $P\text{-value} \approx 0 < 0.01$ we reject H_0. We have convincing evidence that a majority of adult Americans would prefer a hot climate over a cold climate.

10.33
1. p = proportion of all American adults who oppose reinstatement of the draft
2. H_0: $p = 2/3$
3. H_a: $p > 2/3$
4. $\alpha = 0.05$
5. $z = \dfrac{\hat{p} - p}{\sqrt{\dfrac{p(1-p)}{n}}} = \dfrac{\hat{p} - 2/3}{\sqrt{\dfrac{(2/3)(1/3)}{1000}}}$
6. The sample was a random sample from the population. The sample size is much smaller than the population size (the number of American adults). Furthermore, $np = 1000(2/3) = 667 \geq 10$ and $n(1-p) = 1000(1/3) = 333 \geq 10$, so the sample is large enough. Therefore the large sample test is appropriate.
7. $z = \dfrac{700/1000 - 2/3}{\sqrt{\dfrac{(2/3)(1/3)}{1000}}} = 2.23607$
8. $P\text{-value} = P(Z > 2.23607) = 0.01267$
9. Since $P\text{-value} = 0.01267 < 0.05$ we reject H_0. We have convincing evidence that more than two-thirds of American adults oppose reinstatement of the draft.

10.35
1. p = proportion of all cell phone users in 2004 who had received commercial messages or ads
2. H_0: $p = 0.13$
3. H_a: $p > 0.13$
4. $\alpha = 0.05$

5. $z=\dfrac{\hat{p}-p}{\sqrt{\dfrac{p(1-p)}{n}}}=\dfrac{\hat{p}-0.13}{\sqrt{\dfrac{(0.13)(0.87)}{5500}}}$
6. The sample size is much smaller than the population size (the number of cell phone users in 2004). Furthermore, $np=5500(0.13)=715\geq 10$ and $n(1-p)=5500(0.87)=4785\geq 10$, so the sample is large enough. Therefore, if we assume that the sample was a random sample from the population, the large sample test is appropriate.
7. $z=\dfrac{0.2-0.13}{\sqrt{\dfrac{(0.13)(0.87)}{5500}}}=15.436$
8. $P\text{-value}=P(Z>15.436)\approx 0$
9. Since $P\text{-value}\approx 0<0.05$ we reject H_0. We have convincing evidence that the proportion of cell phone users in 2004 who had received commercial messages or ads is more than 0.13.

10.37
1. p = proportion of all adult Americans who believe that playing the lottery would be the best way of accumulating $200,000 in net wealth
2. H_0: $p=0.2$
3. H_a: $p>0.2$
4. $\alpha=0.05$
5. $z=\dfrac{\hat{p}-p}{\sqrt{\dfrac{p(1-p)}{n}}}=\dfrac{\hat{p}-0.2}{\sqrt{\dfrac{(0.2)(0.8)}{1000}}}$
6. We are told to assume that the sample was a random sample from the population. The sample size is much smaller than the population size (the number of adult Americans). Furthermore, $np=1000(0.2)=200\geq 10$ and $n(1-p)=1000(0.8)=800\geq 10$, so the sample is large enough. Therefore the large sample test is appropriate.
7. $z=\dfrac{210/1000-0.2}{\sqrt{\dfrac{(0.2)(0.8)}{1000}}}=0.79057$
8. $P\text{-value}=P(Z>0.79057)=0.21460$
9. Since $P\text{-value}=0.21460>0.05$ we do not reject H_0. We do not have convincing evidence that more than 20% of adult Americans believe that playing the lottery would be the best strategy for accumulating $200,000 in net wealth.

10.39 **a**
1. p = proportion of all adult Americans who believe that the quality of movies being produced is getting worse
2. H_0: $p=0.5$
3. H_a: $p<0.5$
4. $\alpha=0.05$
5. $z=\dfrac{\hat{p}-p}{\sqrt{\dfrac{p(1-p)}{n}}}=\dfrac{\hat{p}-0.5}{\sqrt{\dfrac{(0.5)(0.5)}{1000}}}$
6. The sample was a random sample from the population. The sample size is much smaller than the population size (the number of adult Americans). Furthermore, $np=1000(0.5)=500\geq 10$ and $n(1-p)=1000(0.5)=500\geq 10$, so the sample is large enough. Therefore the large sample test is appropriate.

7. $z = \dfrac{470/1000 - 0.5}{\sqrt{\dfrac{(0.5)(0.5)}{1000}}} = -1.89737$
8. $P\text{-value} = P(Z < -1.89737) = 0.02889$
9. Since $P\text{-value} = 0.02889 < 0.05$ we reject H_0. We have convincing evidence that fewer than half of adult Americans believe that movie quality is decreasing.

b The conditions for performing the test would all still be satisfied. The test statistic would now be

$$z = \frac{47/100 - 0.5}{\sqrt{\dfrac{(0.5)(0.5)}{100}}} = -0.6,$$

which gives $P\text{-value} = P(Z < -0.6) = 0.274 > 0.05$. So in this case, no, we do not have convincing evidence that fewer than half of adult Americans believe that movie quality is decreasing.

c Both results *suggest* that fewer than half of adult Americans believe that movie quality is getting worse. However, getting 470 out of 1000 people responding this way (as opposed to 47 out of 100) provides much *stronger* evidence of this fact.

10.41 The "38%" value given in the article is a proportion of *all* felons; in other words, it is a *population* proportion. Therefore we know that the population proportion is less than 0.4, and there is no need for a hypothesis test.

10.43 **a** $P\text{-value} = 2 \cdot P(t_9 > 0.73) = \mathbf{0.484}$.

b $P\text{-value} = P(t_{10} > -0.5) = \mathbf{0.686}$.

c $P\text{-value} = P(t_{19} < -2.1) = \mathbf{0.025}$.

d $P\text{-value} = P(t_{19} < -5.1) = \mathbf{0.000}$.

e $P\text{-value} = 2 \cdot P(t_{39} > 1.7) = \mathbf{0.097}$.

10.45 **a** $P\text{-value} = P(t_{17} < -2.3) = 0.017 < 0.05$. H_0 is rejected. We have convincing evidence that the mean writing time for all pens of this type is less than 10 hours.

b $P\text{-value} = P(t_{17} < -1.83) = 0.042 > 0.01$. H_0 is not rejected. We do not have convincing evidence that the mean writing time for all pens of this type is less than 10 hours.

c Since t is positive, the sample mean must have been greater than 10. Therefore, we certainly do not have convincing evidence that the mean writing time for all pens of this type is less than 10 hours. H_0 is certainly not rejected.

10.47 **a**
1. μ = mean heart rate after 15 minutes of Wii Bowling for all boys age 10 to 13
2. H_0: $\mu = 98$

3. H_a: $\mu \neq 98$
4. $\alpha = 0.01$
5. $t = \dfrac{\bar{x} - \mu}{s/\sqrt{n}} = \dfrac{\bar{x} - 98}{s/\sqrt{n}}$
6. We are told to assume that it is reasonable to regard the sample of boys as representative of boys age 10 to 13. Under this assumption, it is reasonable to treat the sample as a random sample from the population. We are also told to assume that the distribution of heart rates after 15 minutes of Wii Bowling is approximately normal. So we can proceed with the *t* test.
7. $t = \dfrac{101 - 98}{15/\sqrt{14}} = 0.74833$
8. $P\text{-value} = 2 \cdot P(t_{13} > 0.74833) = 0.468$
9. Since $P\text{-value} = 0.468 > 0.01$ we do not reject H_0. We do not have convincing evidence that the mean heart rate after 15 minutes of Wii Bowling is not equal to 98 beats per minute.

b
1. μ = mean heart rate after 15 minutes of Wii Bowling for all boys age 10 to 13
2. H_0: $\mu = 66$
3. H_a: $\mu > 66$
4. $\alpha = 0.01$
5. $t = \dfrac{\bar{x} - \mu}{s/\sqrt{n}} = \dfrac{\bar{x} - 66}{s/\sqrt{n}}$
6. We are told to assume that it is reasonable to regard the sample of boys as representative of boys age 10 to 13. Under this assumption, it is reasonable to treat the sample as a random sample from the population. We are also told to assume that the distribution of heart rates after 15 minutes of Wii Bowling is approximately normal. So we can proceed with the *t* test.
7. $t = \dfrac{101 - 66}{15/\sqrt{14}} = 8.731$
8. $P\text{-value} = P(t_{13} > 8.731) \approx 0$
9. Since $P\text{-value} \approx 0 < 0.01$ we reject H_0. We have convincing evidence that the mean heart rate after 15 minutes of Wii Bowling is greater than 66 beats per minute.

c It is known that treadmill walking raises the heart rate over the resting heart rate, and the study provided convincing evidence that Wii Bowling does so, also. Although the sample mean heart rate for Wii Bowling was higher than the known population mean heart rate for treadmill walking, the study did not provide convincing evidence of a difference of the population mean heart rate for Wii Bowling from the known population mean for the treadmill.

10.49
1. μ = mean salary offering for accounting graduates at this university
2. H_0: $\mu = 48722$
3. H_a: $\mu > 48722$
4. $\alpha = 0.05$
5. $t = \dfrac{\bar{x} - \mu}{s/\sqrt{n}} = \dfrac{\bar{x} - 48722}{s/\sqrt{n}}$

6. The sample was a random sample from the population. Also, $n = 50 \geq 30$. So we can proceed with the t test.
7. $t = \dfrac{49850 - 48722}{3300/\sqrt{50}} = 2.41702$
8. $P\text{-value} = P(t_{49} > 2.41702) = 0.010$
9. Since $P\text{-value} = 0.010 < 0.05$ we reject H_0. We have convincing evidence that the mean salary offer for accounting graduates of this university is higher than the 2010 national average of $48,722.

10.51
1. μ = mean number of credit cards carried by undergraduates
2. H_0: $\mu = 4.09$
3. H_a: $\mu < 4.09$
4. $\alpha = 0.05$
5. $t = \dfrac{\bar{x} - \mu}{s/\sqrt{n}} = \dfrac{\bar{x} - 4.09}{s/\sqrt{n}}$
6. The sample was a random sample from the population. Also, $n = 132 \geq 30$. So we can proceed with the t test.
7. $t = \dfrac{2.6 - 4.09}{1.2/\sqrt{132}} = -14.266$
8. $P\text{-value} = P(t_{131} < -14.266) \approx 0$
9. Since $P\text{-value} \approx 0 < 0.05$ we reject H_0. We have convincing evidence that the mean number of credit cards carried by undergraduates is less than the credit bureau's figure of 4.09.

10.53
1. μ = mean minimum purchase amount for which Canadians consider it acceptable to use a debit card
2. H_0: $\mu = 10$
3. H_a: $\mu < 10$
4. $\alpha = 0.01$
5. $t = \dfrac{\bar{x} - \mu}{s/\sqrt{n}} = \dfrac{\bar{x} - 10}{s/\sqrt{n}}$
6. The sample was a random sample from the population. Also, $n = 2000 \geq 30$. So we can proceed with the t test.
7. $t = \dfrac{9.15 - 10}{7.6/\sqrt{2000}} = -5.001$
8. $P\text{-value} = P(t_{1999} < -5.001) \approx 0$
9. Since $P\text{-value} \approx 0 < 0.01$ we reject H_0. We have convincing evidence that the mean minimum purchase amount for which Canadians consider it acceptable to use a debit card is less than $10.

10.55 **a**
1. μ = mean weekly time spent using the Internet by Canadians
2. H_0: $\mu = 12.5$
3. H_a: $\mu > 12.5$
4. $\alpha = 0.05$

5. $t = \frac{\bar{x} - \mu}{s/\sqrt{n}} = \frac{\bar{x} - 12.5}{s/\sqrt{n}}$
6. The sample was a random sample from the population. Also, $n = 1000 \geq 30$. So we can proceed with the t test.
7. $t = \frac{12.7 - 12.5}{5/\sqrt{1000}} = 1.26491$
8. $P\text{-value} = P(t_{999} > 1.26491) = 0.103$
9. Since $P\text{-value} = 0.103 > 0.05$ we do not reject H_0. We do not have convincing evidence that the mean weekly time spent using the Internet by Canadians is greater than 12.5 hours.

b Now $t = \frac{12.7 - 12.5}{2/\sqrt{1000}} = 3.16228$, which gives $P\text{-value} = P(t_{999} > 3.16228) = 0.001$. Since $P\text{-value} = 0.001 < 0.05$ we reject H_0. We have convincing evidence that the mean weekly time spent using the Internet by Canadians is greater than 12.5 hours.

c The sample standard deviation of 2 in Part (b) means that the population of weekly Internet times has a standard deviation of around 2. Likewise, the sample standard deviation of 5 in Part (a) means that the population of weekly Internet times has a standard deviation of around 5. Assuming that the population of weekly Internet times has a mean of 12.5, it is far less likely to get a sample mean of 12.7 if the population standard deviation is 2 than if the population standard deviation is 5, since greater deviations from the mean are expected when the population standard deviation is larger. This explains why H_0 is rejected when the sample standard deviation is 2, but not when the sample standard deviation is 5.

10.57 **a** Yes. Since the pattern in the normal probability plot is roughly linear, and since the sample was a random sample from the population, the t test is appropriate.

b The boxplot shows a median of around 245, and since the distribution is roughly symmetrical distribution, this tells us that the sample mean is around 245, also. This might initially suggest that the population mean differs from 240. But when you consider the fact that the sample is relatively small, and that the sample values range all the way from 225 to 265, you realize that such a sample mean would still be feasible if the population mean were 240.

c
1. μ = mean calorie content for frozen dinners of this type
2. H_0: $\mu = 240$
3. H_a: $\mu \neq 240$
4. $\alpha = 0.05$
5. $t = \frac{\bar{x} - \mu}{s/\sqrt{n}} = \frac{\bar{x} - 240}{s/\sqrt{n}}$
6. As explained in Part (a), the conditions for performing the t test are met.
7. The mean and standard deviation of the sample values are 244.33333 and 12.38278, respectively. So $t = \frac{244.33333 - 240}{12.38278/\sqrt{12}} = 1.21226$.
8. $P\text{-value} = 2 \cdot P(t_{11} > 1.21226) = 0.251$
9. Since $P\text{-value} = 0.251 > 0.05$ we do not reject H_0. We do not have convincing evidence that the mean calorie content for frozen dinners of this type differs from 240.

10.59 **a** Increasing the sample size increases the power.

b Increasing the significance level increases the power.

10.61 **a** $\alpha =$ area under standard normal curve to the left of $-1.28 =$ **0.1**.

b When $z = -1.28$, $\bar{x} = 10 + (-1.28)(.1) = 9.872$. So H_0 is rejected for values of $\bar{x} \le 9.872$.
If $\mu = 9.8$, then $\bar{x}$ is normally distributed with mean 9.8 and standard deviation 0.1. So
$P(H_0$ is rejected$) = P(\bar{x} \le 9.872)$
= area under standard normal curve to left of $(9.872 - 9.8)/0.1$
= area under standard normal curve to left of 0.72
$= 0.7642$.
So $\beta = P(H_0$ not rejected$) = 1 - 0.7642 =$ **0.2358**.

c H_0 states that $\mu = 10$ and H_a states that $\mu < 10$. Since 9.5 is further from 10 (in the direction indicated by H_a), β is **less** for $\mu = 9.5$ than for $\mu = 9.8$.
For $\mu = 9.5$,
$P(H_0$ is rejected$) = P(\bar{x} \le 9.872)$
= area under standard normal curve to left of $(9.872 - 9.5)/0.1$
= area under standard normal curve to left of 3.72
$= 0.9999$.
So $\beta = P(H_0$ not rejected$) = 1 - 0.9999 =$ **0.0001**.

d Power when $\mu = 9.8$ is $1 - 0.2358 =$ **0.7642**.
Power when $\mu = 9.5$ is $1 - 0.0001 =$ **0.9999**.

10.63 **a** $t = \dfrac{0.0372 - 0.035}{0.0125/\sqrt{7}} = 0.46565$. So P-value $= P(t_6 > 0.46565) = 0.329 > 0.05$. Therefore, H_0 is not rejected, and we do not have convincing evidence that $\mu > 0.035$.

b $d = \dfrac{|0.04 - 0.035|}{0.0125} = 0.4$. Using Appendix Table 5, for a one-tailed test, $\alpha = 0.05$, 6 degrees of freedom, we get $\beta \approx$ **0.75**.

c Power $\approx 1 - 0.75 =$ **0.25**.

10.65 Using Appendix Table 5:

a $d = \dfrac{|0.52 - 0.5|}{0.02} = 1$, $\beta \approx$ **0.04**.

b $d = \dfrac{|0.48 - 0.5|}{0.02} = 1$, $\beta \approx$ **0.04**.

c $d = \frac{|0.52 - 0.5|}{0.02} = 1, \ \beta \approx \mathbf{0.24}.$

d $d = \frac{|0.54 - 0.5|}{0.02} = 2, \ \beta \approx \mathbf{0}.$

e $d = \frac{|0.54 - 0.5|}{0.04} = 1, \ \beta \approx \mathbf{0.04}.$

f $d = \frac{|0.54 - 0.5|}{0.04} = 1, \ \beta \approx \mathbf{0.01}.$

g Comparing Part (b) with Part (a), it makes sense that true values of μ equal distances to the right and left of the hypothesized value will give equal probabilities of a Type II error.
Comparing Part (c) with Part (a), it makes sense that a smaller significance level will give a larger probability of a Type II error (since with a smaller significance level you are less likely to reject H_0, and therefore more likely to fail to reject H_0).
Comparing Part (d) with Part (a), it makes sense that an alternative value of μ further from the hypothesized value will give a smaller probability of a Type II error (since the test is more likely to correctly detect a true value of μ that is further from the hypothesized value of μ).
Comparing Part (e) with Part (a), it makes sense that an alternative value of μ twice as far from the hypothesized value combined with a population standard deviation that is twice as large will give the same probability of a Type II error.
Comparing Part (f) with Part (e), it makes sense that a larger sample size will give a smaller probability of a Type II error (since a larger sample is more likely to detect that the true value of μ is not equal to the hypothesized value of μ).

10.67 a

1. p = proportion of all women who would like to choose a baby's sex who would choose a girl.
2. H_0: $p = 0.5$
3. H_a: $p \neq 0.5$
4. $\alpha = 0.05$
5. $z = \frac{\hat{p} - p}{\sqrt{\frac{p(1-p)}{n}}} = \frac{\hat{p} - 0.5}{\sqrt{\frac{(0.5)(0.5)}{229}}}$
6. We need to assume that the sample was a random sample from the population of women who would like to choose the sex of a baby. The sample size is presumably much smaller than the population size (the number of women who would like to choose the sex of a baby). Also, $np = 229(0.5) = 114.5 \geq 10$ and $n(1-p) = 229(0.5) = 114.5 \geq 10$, so the sample is large enough. Therefore the large sample test is appropriate.
7. $z = \frac{140/229 - 0.5}{\sqrt{\frac{(0.5)(0.5)}{229}}} = 3.370$
8. P-value $= P(Z > 3.370) = 0.0004$

9. Since $P\text{-value} = 0.0004 < 0.05$ we reject H_0. We have convincing evidence that the proportion of all women who would like to choose a baby's sex who would choose a girl is not equal to 0.5. (This contradicts the statement in the article.)

b The survey was conducted only on women who had visited the Center for Reproductive Medicine at Brigham and Women's Hospital. It is quite possible that women who choose this institution have views on the matter that are different from those of women in general. (This is selection bias.) Also, with only 561 of the 1385 women responding, it is quite possible that the majority who did not respond had different views from those who did. (This is nonresponse bias.) For these two reasons it seems unreasonable to generalize the results to a larger population.

10.69
1. p = proportion of all U.S. adults who believe that rudeness is a worsening problem
2. H_0: $p = 0.75$
3. H_a: $p < 0.75$
4. $\alpha = 0.05$
5. $z = \dfrac{\hat{p} - p}{\sqrt{\dfrac{p(1-p)}{n}}} = \dfrac{\hat{p} - 0.75}{\sqrt{\dfrac{(0.75)(0.25)}{2013}}}$
6. We need to assume that the sample was a random sample from the population of U.S. adults. The sample size is much smaller than the population size (the number of U.S. adults). Furthermore, $np = 2013(0.75) = 1509.75 \geq 10$ and $n(1-p) = 2013(0.25) = 503.25 \geq 10$, so the sample is large enough. Therefore the large sample test is appropriate.
7. $z = \dfrac{1283/2013 - 0.75}{\sqrt{\dfrac{(0.75)(0.25)}{2013}}} = -11.671$
8. $P\text{-value} = P(Z < -11.671) \approx 0$
9. Since $P\text{-value} \approx 0 < 0.05$ we reject H_0. We have convincing evidence that less than three-quarters of all U.S. adults believe that rudeness is a worsening problem.

10.71
1. μ = mean time to distraction for Australian teenage boys
2. H_0: $\mu = 5$
3. H_a: $\mu < 5$
4. $\alpha = 0.01$
5. $t = \dfrac{\bar{x} - \mu}{s/\sqrt{n}} = \dfrac{\bar{x} - 5}{s/\sqrt{n}}$
6. We are told to assume that the sample was a random sample from the population. Also, $n = 50 \geq 30$. So we can proceed with the t test.
7. $t = \dfrac{4 - 5}{1.4/\sqrt{50}} = -5.051$
8. $P\text{-value} = P(t_{49} < -5.051) \approx 0$
9. Since $P\text{-value} \approx 0 < 0.01$ we reject H_0. We have convincing evidence that the mean time to distraction for Australian teenage boys is less than 5 minutes.

10.73
1. p = proportion of all U.S. adults who approve of casino gambling
2. H_0: $p = 2/3$
3. H_a: $p > 2/3$

4. $\alpha = 0.05$
5. $z = \dfrac{\hat{p} - p}{\sqrt{\dfrac{p(1-p)}{n}}} = \dfrac{\hat{p} - 2/3}{\sqrt{\dfrac{(2/3)(1/3)}{1523}}}$
6. We need to assume that the sample selected at random from households with telephones was a random sample from the population of U.S. adults. The sample size is much smaller than the population size (the number of U.S. adults). Furthermore, $np = 1523(2/3) = 1015 \geq 10$ and $n(1-p) = 1523(1/3) = 508 \geq 10$, so the sample is large enough. Therefore the large sample test is appropriate.
7. $z = \dfrac{1035/1523 - 2/3}{\sqrt{\dfrac{(2/3)(1/3)}{1523}}} = 1.06902$
8. $P\text{-value} = P(Z > 1.06902) = 0.143$
9. Since $P\text{-value} = 0.143 > 0.05$ we do not reject H_0. We do not have convincing evidence that more than two-thirds of all U.S. adults approve of casino gambling.

10.75
1. p = proportion of all U.S. adults who believe that an investment of $25 per week over 40 years with a 7% annual return would result in a sum of over $100,000
2. H_0: $p = 0.4$
3. H_a: $p < 0.4$
4. $\alpha = 0.05$
5. $z = \dfrac{\hat{p} - p}{\sqrt{\dfrac{p(1-p)}{n}}} = \dfrac{\hat{p} - 0.4}{\sqrt{\dfrac{(0.4)(0.6)}{1010}}}$
6. The sample was random sample from the population of U.S. adults. The sample size is much smaller than the population size (the number of U.S. adults). Furthermore, $np = 1010(0.4) = 404 \geq 10$ and $n(1-p) = 1010(0.6) = 606 \geq 10$, so the sample is large enough. Therefore the large sample test is appropriate.
7. $z = \dfrac{374/1010 - 0.4}{\sqrt{\dfrac{(0.4)(0.6)}{1010}}} = -1.92688$
8. $P\text{-value} = P(Z < -1.92688) = 0.027$
9. Since $P\text{-value} = 0.027 < 0.05$ we reject H_0. We have convincing evidence that less than 40% of all U.S. adults believe that an investment of $25 per week over 40 years with a 7% annual return would result in a sum of over $100,000.

10.77 **a**
1. μ = mean weight for non-top-20 starters
2. H_0: $\mu = 105$
3. H_a: $\mu < 105$
4. $\alpha = 0.05$
5. $t = \dfrac{\bar{x} - \mu}{s/\sqrt{n}} = \dfrac{\bar{x} - 105}{s/\sqrt{n}}$
6. The sample was a random sample from the population. Also, $n = 33 \geq 30$. So we can proceed with the t test.

7. $t = \dfrac{103.3 - 105}{16.3/\sqrt{33}} = -0.59913$
8. P-value $= P(t_{32} < -0.59913) = 0.277$
9. Since $P\text{-value} = 0.277 > 0.01$ we do not reject H_0. We do not have convincing evidence that the mean weight for non-top-20 starters is less than 105 kg.

10.79
1. p = proportion of all people who would respond if the distributor is fitted with an eye patch
2. H_0: $p = 0.4$
3. H_a: $p > 0.4$
4. $\alpha = 0.05$
5. $z = \dfrac{\hat{p} - p}{\sqrt{\dfrac{p(1-p)}{n}}} = \dfrac{\hat{p} - 0.4}{\sqrt{\dfrac{(0.4)(0.6)}{200}}}$
6. We have to assume that the sample was random sample from the population of people who could be approached with a questionnaire. The sample size is much smaller than the population size. Furthermore, $np = 200(0.4) = 80 \geq 10$ and $n(1-p) = 200(0.6) = 120 \geq 10$, so the sample is large enough. Therefore the large sample test is appropriate.
7. $z = \dfrac{109/200 - 0.4}{\sqrt{\dfrac{(0.4)(0.6)}{200}}} = 4.186$
8. P-value $= P(Z > 4.186) \approx 0$
9. Since $P\text{-value} \approx 0 < 0.05$ we reject H_0. We have convincing evidence that more than 40% of all people who could be approached with a questionnaire will respond when the distributor wears an eye patch.

10.81
1. μ = mean daily revenue since the change
2. H_0: $\mu = 75$
3. H_a: $\mu < 75$
4. $\alpha = 0.05$
5. $t = \dfrac{\bar{x} - \mu}{s/\sqrt{n}} = \dfrac{\bar{x} - 75}{s/\sqrt{n}}$
6. The sample was a random sample of days. In order to proceed with the t test we must assume that the distribution of daily revenues since the change is normal.
7. $t = \dfrac{70 - 75}{4.2/\sqrt{20}} = -5.324$
8. P-value $= P(t_{19} < -5.324) \approx 0$
9. Since $P\text{-value} \approx 0 < 0.05$ we reject H_0. We have convincing evidence that the mean daily revenue has decreased since the price increase.

Cumulative Review Exercises

CR10.1

Gather a set of volunteer older people with knee osteoarthritis (we will assume that 40 such volunteers are available). Have each person rate his/her knee pain on a scale of 1-10, where 10 is the worst pain. Randomly assign the volunteers to two groups, A and B, of equal sizes. (This can be done by writing the names of the volunteers onto slips of paper. Place the slips into a hat, and pick 20 at random. These 20 people will go into Group A, and the remaining 20 people will go into Group B.) The volunteers in Group A will attend twice weekly sessions of one hour of tai chi. The volunteers in Group B will simply continue with their lives as they usually would. After 12 weeks, each volunteer should be asked to rate his/her pain on the same scale as before. The mean reduction in pain for Group A should then be compared to the mean reduction in pain for Group B.

CR10.3

a

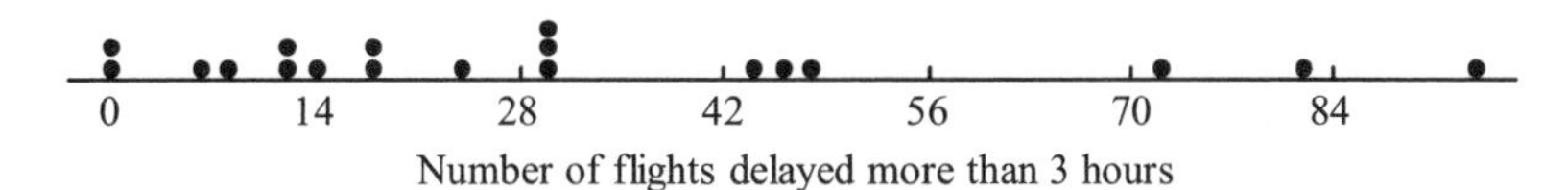

There are three airlines that stand out from the rest by having large numbers of delayed flights. These airlines are ExpressJet, Delta, and Continental, with 93, 81, and 72 delayed flights, respectively.

b

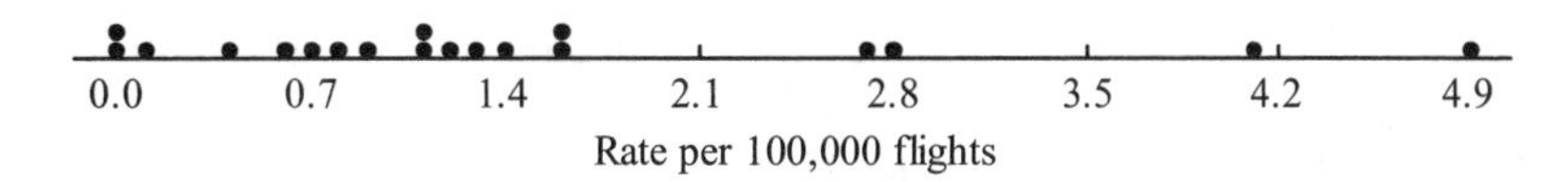

A typical number of flights delayed per 10,000 flights is around 1.1, with most rates lying between 0 and 1.6. There are four airlines that standout from the rest by having particularly high rates, with two of those four having *particularly* high rates.

c The rate per 100,000 flights data should be used, since this measures the likelihood of any given flight being late. An airline could standout in the number of flights delayed data purely as a result of having a large number of flights.

CR10.5

a The number of people in the sample who change their passwords quarterly is binomially distributed with $n = 20$ and $p = 0.25$. So, using Appendix Table 9, $p(3) = \mathbf{0.134}$.

b Using Appendix Table 9, $P(\text{more than 8 change passwords quarterly})$
$= 0.027 + 0.010 + 0.003 + 0.001 = \mathbf{0.041}$.

c $\mu_x = np = 100(0.25) = \mathbf{25}$, $\sigma_x = \sqrt{np(1-p)} = \sqrt{100(0.25)(0.75)} = \mathbf{4.330}$.

d Since $np = 100(0.25) = 25 \geq 10$ and $n(1-p) = 100(0.75) = 75 \geq 10$ the normal approximation to the binomial distribution can be used. Thus,

$$P(x < 20) \approx P\left(z \leq \frac{19.5 - 25}{4.33013}\right) = P(z \leq -1.27017) = \mathbf{0.102}.$$

CR10.7

a $P(O) = \mathbf{0.4}$.

b Anyone who accepts a job offer must have received at least one job offer, so $P(A) = P(O \cap A) = P(A \mid O)P(O) = (0.45)(0.4) = \mathbf{0.18}$.

c $P(G) = \mathbf{0.26}$.

d $P(A \mid O) = \mathbf{0.45}$.

e Since anyone who accepts a job offer must have received at least one job offer, $P(O \mid A) = \mathbf{1}$.

f $P(A \cap O) = P(A) = \mathbf{0.18}$.

CR10.9

a Check of Conditions

1. Since $n\hat{p} = 115(38/115) = 38 \geq 10$ and $n(1-\hat{p}) = 2002(77/2002) = 77 \geq 10$, the sample size is large enough.
2. The sample size of $n = 115$ is much smaller than 10% of the population size (the number of U.S. medical residents).
3. We are told to regard the sample as a random sample from the population.

Calculation

The 95% confidence interval for p is

$$\hat{p} \pm 1.96\sqrt{\frac{\hat{p}(1-\hat{p})}{n}} = \frac{38}{115} \pm 1.96\sqrt{\frac{(38/115)(77/115)}{115}} = (\mathbf{0.244, 0.416}).$$

Interpretation

We are 95% confident that the proportion of all U.S. medical residents who work moonlighting jobs is between 0.244 and 0.416.

b Check of Conditions

1. Since $n\hat{p} = 115(22/115) = 22 \geq 10$ and $n(1-\hat{p}) = 115(93/2002) = 93 \geq 10$, the sample size is large enough.
2. The sample size of $n = 115$ is much smaller than 10% of the population size (the number of U.S. medical residents).
3. We are told to regard the sample as a random sample from the population.

Calculation

The 90% confidence interval for p is

$$\hat{p} \pm 1.645\sqrt{\frac{\hat{p}(1-\hat{p})}{n}} = \frac{22}{115} \pm 1.645\sqrt{\frac{(22/115)(93/115)}{115}} = (\mathbf{0.131, 0.252}).$$

Interpretation

We are 90% confident that the proportion of all U.S. medical residents who have credit card debt of more than \$3000 is between 0.131 and 0.252.

c The interval in Part (a) is wider than the interval in Part (b) because the confidence level in Part (a) (95%) is greater than the confidence level in Part (b) (90%) and because the sample proportion in Part (a) (38/115) is closer to 0.5 than the sample proportion in Part (b) (22/115).

CR10.11

A reasonable estimate of σ is given by $(\text{sample range})/4 = (20.3 - 19.9)/4 = 0.1$. Thus

$$n = \left(\frac{1.96\sigma}{B}\right)^2 = \left(\frac{1.96 \cdot 0.1}{0.01}\right)^2 = 384.16.$$

So we need a sample size of **385**.

CR10.13

1. p = proportion of all baseball fans who believe that the designated hitter rule should either be expanded to both baseball leagues or eliminated
2. H_0: $p = 0.5$
3. H_a: $p > 0.5$
4. $\alpha = 0.05$
5. $z = \dfrac{\hat{p} - p}{\sqrt{\dfrac{p(1-p)}{n}}} = \dfrac{\hat{p} - 0.5}{\sqrt{\dfrac{(0.5)(0.5)}{394}}}$
6. The sample was a random sample from the population. The sample size is much smaller than the population size (the number of baseball fans). Furthermore, $np = 394(0.5) = 197 \geq 10$ and $n(1-p) = 394(0.5) = 197 \geq 10$, so the sample is large enough. Therefore the large sample test is appropriate.
7. $z = \dfrac{272/394 - 0.5}{\sqrt{\dfrac{(0.5)(0.5)}{394}}} = 7.557$
8. $P\text{-value} = P(Z > 7.557) \approx 0$
9. Since $P\text{-value} \approx 0 < 0.05$ we reject H_0. We have convincing evidence that a majority of baseball fans believe that the designated hitter rule should either be expanded to both baseball leagues or eliminated.

CR10.15

a With a sample mean of 14.6, the sample standard deviation of 11.6 places zero just over one standard deviation below the mean. Since no teenager can spend a negative time online, to get a typical deviation from the mean of just over 1, there must be values that are substantially more than one standard deviation above the mean. This suggests that the distribution of online times in the sample is positively skewed.

b

1. μ = mean weekly time online for teenagers
2. H_0: $\mu = 10$
3. H_a: $\mu > 10$
4. $\alpha = 0.05$
5. $t = \dfrac{\overline{x} - \mu}{s/\sqrt{n}} = \dfrac{\overline{x} - 10}{s/\sqrt{n}}$
6. The sample was a random sample of teenagers. Also, $n = 534 \geq 30$. Therefore we can proceed with the t test.

7. $t = \dfrac{14.6 - 10}{11.6/\sqrt{534}} = 9.164.$
8. $P\text{-value} = P(t_{533} > 9.164) \approx 0$
9. Since $P\text{-value} \approx 0 < 0.05$ we reject H_0. We have convincing evidence that the mean weekly time online for teenagers is greater than 10 hours.

Chapter 11
Comparing Two Populations or Treatments

Note: In this chapter, numerical answers to questions involving the normal and t distributions were found using values from a calculator. Students using statistical tables will find that their answers differ slightly from those given.

11.1 Since n_1 and n_2 are large, the distribution of $\bar{x}_1 - \bar{x}_2$ is approximately normal. Its mean is $\mu_1 - \mu_2 = 30 - 25 = \mathbf{5}$ and its standard deviation is $\sqrt{\frac{\sigma_1^2}{n_1} + \frac{\sigma_2^2}{n_2}} = \sqrt{\frac{2^2}{40} + \frac{3^2}{50}} = \mathbf{0.529}$.

11.3 **a** We need to assume that the 22 heart attack patients who were dog owners formed a random sample from the set of all heart attack patients who are dog owners and that the 80 heart attack patients who did not own a dog formed an independent random sample from the set of all heart attack patients who do not own a dog. Also, since the sample of size 22 is not large, we need to assume that the distribution of the HRVs of all heart attack patients who are dog owners is normal.

b
1. μ_1 = mean HRV for all heart attack patients who are dog owners
   μ_2 = mean HRV for all heart attack patients who do not own a dog
2. H_0: $\mu_1 - \mu_2 = 0$
3. H_a: $\mu_1 - \mu_2 \neq 0$
4. $\alpha = 0.05$
5. $t = \dfrac{(\bar{x}_1 - \bar{x}_2) - (\text{hypothesized value})}{\sqrt{\frac{s_1^2}{n_1} + \frac{s_2^2}{n_2}}} = \dfrac{(\bar{x}_1 - \bar{x}_2) - 0}{\sqrt{\frac{s_1^2}{n_1} + \frac{s_2^2}{n_2}}}$
6. As stated in Part (a), we need to assume that the 22 heart attack patients who were dog owners formed a random sample from the set of all heart attack patients who are dog owners and that the 80 heart attack patients who did not own a dog formed an independent random sample from the set of all heart attack patients who do not own a dog, and that the distributions of the HRVs of all heart attack patients who are dog owners and of all heart attack patients who do not own dogs are normal.
7. $t = \dfrac{873 - 800}{\sqrt{\frac{136^2}{22} + \frac{134^2}{80}}} = 2.237$
8. df = 33.083
   $P\text{-value} = 2 \cdot P(t_{33.083} > 2.23672) = 0.032$
9. Since $P\text{-value} = 0.032 < 0.05$ we reject H_0. We have convincing evidence that the mean HRV for all heart attack patients who are dog owners is not equal to the mean HRV for all heart attack patients who do not own a dog. This conclusion is consistent with that of the paper.

11.5 **a**

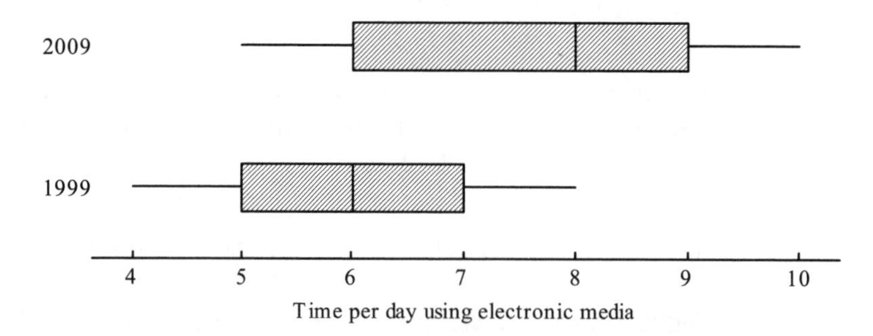

We need to assume that the population distributions of time per day using electronic media are normal. Since the boxplots are roughly symmetrical and since there is no outlier in either sample this assumption is justified, and it is therefore reasonable to carry out a two-sample t test.

b 1. μ_1 = mean time using electronic media for all kids age 8 to 18 in 2009
μ_2 = mean time using electronic media for all kids age 8 to 18 in 1999

2. H_0: $\mu_1 - \mu_2 = 0$
3. H_a: $\mu_1 - \mu_2 > 0$
4. $\alpha = 0.01$
5. $t = \dfrac{(\bar{x}_1 - \bar{x}_2) - (\text{hypothesized value})}{\sqrt{\dfrac{s_1^2}{n_1} + \dfrac{s_2^2}{n_2}}} = \dfrac{(\bar{x}_1 - \bar{x}_2) - 0}{\sqrt{\dfrac{s_1^2}{n_1} + \dfrac{s_2^2}{n_2}}}$
6. We are told to assume that it is reasonable to regard the two samples as representative of kids age 8 to 18 in each of the two years when the surveys were conducted. We can then treat the samples as random samples from their respective populations. Also, as discussed in Part (a), the boxplots show that it is reasonable to assume that the population distributions are normal. So we can proceed with a two-sample t test.
7. $\bar{x}_1 = 7.6 \quad s_1 = 1.595 \quad \bar{x}_2 = 5.933 \quad s_2 = 1.100$

$$t = \frac{7.6 - 5.933}{\sqrt{\dfrac{1.595^2}{15} + \dfrac{1.100^2}{15}}} = 3.332$$

8. df = 24.861
$P\text{-value} = P(t_{24.861} > 3.332) = 0.001$
9. Since $P\text{-value} = 0.001 < 0.01$ we reject H_0. We have convincing evidence that the mean number of hours per day spent using electronic media was greater in 2009 than in 1999.

c As explained in Parts (a) and (b), the conditions for the two-sample t test or interval are satisfied. A 98% confidence interval for $\mu_1 - \mu_2$ is

$$(\bar{x}_1 - \bar{x}_2) \pm (t \text{ critical value})\sqrt{\frac{s_1^2}{n_1} + \frac{s_2^2}{n_2}}$$
$$= (7.6 - 5.93333) \pm 2.48605\sqrt{\frac{1.595^2}{15} + \frac{1.100^2}{15}}$$
$$= (\mathbf{0.423, 2.910})$$

We are 98% confident that the difference between the mean number of hours per day spent using electronic media in 2009 and 1999 is between 0.423 and 2.910.

11.7

1. μ_1 = mean food intake for the 4-hour sleep treatment
   μ_2 = mean food intake for the 8-hour sleep treatment
2. H_0: $\mu_1 - \mu_2 = 0$
3. H_a: $\mu_1 - \mu_2 \neq 0$
4. $\alpha = 0.05$
5. $t = \dfrac{(\bar{x}_1 - \bar{x}_2) - (\text{hypothesized value})}{\sqrt{\dfrac{s_1^2}{n_1} + \dfrac{s_2^2}{n_2}}} = \dfrac{(\bar{x}_1 - \bar{x}_2) - 0}{\sqrt{\dfrac{s_1^2}{n_1} + \dfrac{s_2^2}{n_2}}}$
6.

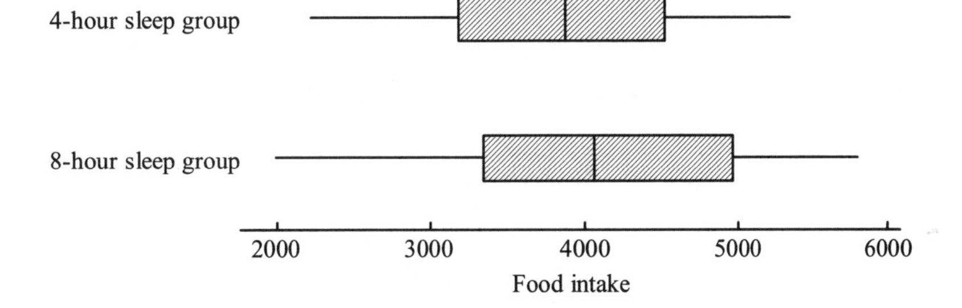

The experimental subjects were randomly assigned to the two sleep conditions. Also, since the two boxplots are roughly symmetrical and there was no outlier in either group we are justified in assuming that the food intake distributions for the two treatments are normal. Thus, we can proceed with the two-sample t test.

7. $\bar{x}_1 = 3924 \quad s_1 = 829.668 \quad \bar{x}_2 = 4069.267 \quad s_2 = 952.896$
   $$t = \frac{3924 - 4069.267}{\sqrt{\frac{829.668^2}{15} + \frac{952.896^2}{15}}} = -0.445$$
8. df = 27.480
   $P\text{-value} = 2 \cdot P(t_{27.480} < -0.445) = 0.660$
9. Since $P\text{-value} = 0.660 > 0.05$ we do not reject H_0. We do not have convincing evidence of a difference in the means for the two sleep treatments.

11.9 **a** If the vertebroplasty group had been compared to a group of patients who did not receive any treatment, and if, for example, the people in the vertebroplasty group experienced a greater pain reduction on average than the people in the "no treatment" group, then it would be impossible to tell whether the observed pain reduction in the vertebroplasty group was caused by the treatment or merely by the subjects' knowledge that some treatment was being applied.

By using a placebo group it is ensured that the subjects in both groups have the knowledge of some "treatment," so that any differences between the pain reduction in the two groups can be attributed to the nature of the vertebroplasty treatment.

b Check of Conditions

Since $n_1 = 68 \geq 30$ and $n_2 = 63 \geq 30$, if we assume that the subjects were randomly assigned to the treatments, we can proceed with construction of a two-sample t interval.

Calculation

df = 127.402. The 95% confidence interval for $\mu_1 - \mu_2$ is

$$(\bar{x}_1 - \bar{x}_2) \pm (t \text{ critical value})\sqrt{\frac{s_1^2}{n_1} + \frac{s_2^2}{n_2}}$$
$$= (4.2 - 3.9) \pm 1.979\sqrt{\frac{2.8^2}{68} + \frac{2.9^2}{63}}$$
$$= (\mathbf{-0.687, 1.287})$$

Interpretation

We are 95% confident that the difference in mean pain intensity 3 days after treatment for the vertebroplasty treatment and the fake treatment is between −0.687 and 1.287.

c 14 days:

Check of Conditions

See Part (b).

Calculation

df = 128.774. The 95% confidence interval for $\mu_1 - \mu_2$ is

$$(\bar{x}_1 - \bar{x}_2) \pm (t \text{ critical value})\sqrt{\frac{s_1^2}{n_1} + \frac{s_2^2}{n_2}}$$
$$= (4.3 - 4.5) \pm 1.979\sqrt{\frac{2.9^2}{68} + \frac{2.8^2}{63}}$$
$$= (\mathbf{-1.186, 0.786})$$

Interpretation

We are 95% confident that the difference in mean pain intensity 14 days after treatment for the vertebroplasty treatment and the fake treatment is between −1.186 and 0.786.

1 month:

Check of Conditions

See Part (b).

Calculation

df = 127.435. The 95% confidence interval for $\mu_1 - \mu_2$ is

$$(\bar{x}_1 - \bar{x}_2) \pm (t \text{ critical value})\sqrt{\frac{s_1^2}{n_1} + \frac{s_2^2}{n_2}}$$
$$= (3.9 - 4.6) \pm 1.979\sqrt{\frac{2.9^2}{68} + \frac{3.0^2}{63}}$$
$$= (\mathbf{-1.722, 0.322})$$

Interpretation
We are 95% confident that the difference in mean pain intensity 1 month after treatment for the vertebroplasty treatment and the fake treatment is between −1.722 and 0.322.

d The fact that all of the intervals contain zero tells us that we do not have convincing evidence at the 0.05 level of a difference in the mean pain intensity for the vertebroplasty treatment and the fake treatment at any of the three times.

11.11
1. μ_1 = mean daily commute for Calgary working males
   μ_2 = mean daily commute for Calgary working females
2. H_0: $\mu_1 - \mu_2 = 0$
3. H_a: $\mu_1 - \mu_2 \neq 0$
4. $\alpha = 0.05$
5. $t = \dfrac{(\bar{x}_1 - \bar{x}_2) - (\text{hypothesized value})}{\sqrt{\dfrac{s_1^2}{n_1} + \dfrac{s_2^2}{n_2}}} = \dfrac{(\bar{x}_1 - \bar{x}_2) - 0}{\sqrt{\dfrac{s_1^2}{n_1} + \dfrac{s_2^2}{n_2}}}$
6. We are told that the samples were random samples from the populations. Also $n_1 = 247 \geq 30$ and $n_2 = 253 \geq 30$, so we can proceed with the two-sample *t* test.
7. $t = \dfrac{29.6 - 27.3}{\sqrt{\dfrac{24.3^2}{247} + \dfrac{24.0^2}{253}}} = 1.065$
8. df = 497.339
   $P\text{-value} = 2 \cdot P(t_{497.339} > 1.065) = 0.288$
9. Since $P\text{-value} = 0.288 > 0.05$ we do not reject H_0. We do not have convincing evidence that the mean commute times for male and female working Calgary residents differ.

11.13 **a** μ_1 = mean payment for claims not involving errors
μ_2 = mean payment for claims involving errors
H_0: $\mu_1 - \mu_2 = 0$
H_a: $\mu_1 - \mu_2 < 0$

b Answer: **(ii) 2.65**. Since the samples are large, we are using a *t* distribution with a large number of degrees of freedom, which can be approximated with the standard normal distribution. $P(Z > 2.65) = 0.004$, which is the *P*-value given. None of the other possible values of *t* gives the correct *P*-value.

11.15 **a** Check of Conditions

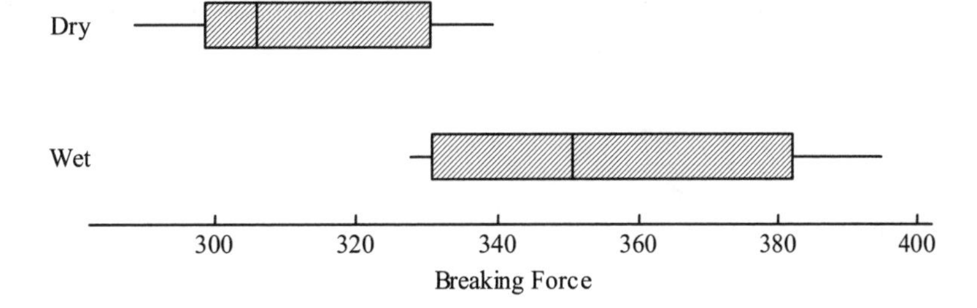

The boxplots are roughly symmetrical (in such small samples no greater degree of symmetry can be expected, even from perfectly normally distributed populations) and neither data set contains outliers, so we are justified in assuming normal distributions for the populations. Therefore, if we assume that the cement bonds were randomly assigned to the treatments, we can proceed with construction of a two-sample t interval.

Calculation

df = 8.765. The 90% confidence interval for $\mu_1 - \mu_2$ is

$$(\bar{x}_1 - \bar{x}_2) \pm (t \text{ critical value})\sqrt{\frac{s_1^2}{n_1} + \frac{s_2^2}{n_2}}$$

$$= (311.6 - 355.583) \pm 1.839\sqrt{\frac{18.377^2}{6} + \frac{27.270^2}{6}}$$

$$= (\mathbf{-68.668, -19.299})$$

Interpretation

We are 90% confident that the difference between the mean breaking force in a dry medium at 37 degrees and the mean breaking force at the same temperature in a wet medium is between −68.668 and −19.299.

b 1. μ_1 = mean breaking strength in a dry medium at 37 degrees
μ_2 = mean breaking strength in a dry medium at 22 degrees

2. H_0: $\mu_1 - \mu_2 = 100$
3. H_a: $\mu_1 - \mu_2 > 100$
4. $\alpha = 0.1$
5. $$t = \frac{(\bar{x}_1 - \bar{x}_2) - (\text{hypothesized value})}{\sqrt{\frac{s_1^2}{n_1} + \frac{s_2^2}{n_2}}} = \frac{(\bar{x}_1 - \bar{x}_2) - 100}{\sqrt{\frac{s_1^2}{n_1} + \frac{s_2^2}{n_2}}}$$
6.

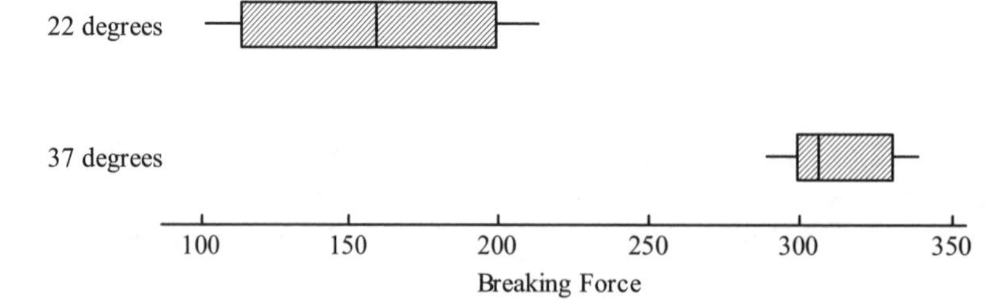

The boxplots are roughly symmetrical and neither data set contains outliers, so we are justified in assuming normal distributions for the populations. Therefore, if we assume that the cement bonds were randomly assigned to the treatments, we can proceed with the two-sample t test.

7. $$t = \frac{(311.6 - 157.517) - (100)}{\sqrt{\frac{18.377^2}{6} + \frac{44.307^2}{6}}} = 2.762$$
8. df = 6.671
$P\text{-value} = P(t_{6.671} > 2.762) = 0.015$

9. Since $P\text{-value} = 0.015 < 0.1$ we reject H_0. We have convincing evidence that the mean breaking force in a dry medium at the higher temperature is greater than the mean breaking force at the lower temperature by more than 100 N.

11.17 a

1. μ_1 = mean percentage of time playing with police car for male monkeys
   μ_2 = mean percentage of time playing with police car for female monkeys
2. H_0: $\mu_1 - \mu_2 = 0$
3. H_a: $\mu_1 - \mu_2 > 0$
4. $\alpha = 0.05$
5. $t = \dfrac{(\bar{x}_1 - \bar{x}_2) - (\text{hypothesized value})}{\sqrt{\dfrac{s_1^2}{n_1} + \dfrac{s_2^2}{n_2}}} = \dfrac{(\bar{x}_1 - \bar{x}_2) - 0}{\sqrt{\dfrac{s_1^2}{n_1} + \dfrac{s_2^2}{n_2}}}$
6. We are told that that it is reasonable to regard these two samples of 44 monkeys as representative of the populations of male and female monkeys. It is therefore reasonable to regard them as random samples. Also $n_1 = 44 \geq 30$ and $n_2 = 44 \geq 30$, so we can proceed with the two-sample t test.
7. $t = \dfrac{18 - 8}{\sqrt{\dfrac{5^2}{44} + \dfrac{4^2}{44}}} = 10.359$
8. $\text{df} = 82.047$
   $P\text{-value} = P(t_{82.047} > 10.359) \approx 0$
9. Since $P\text{-value} \approx 0 < 0.05$ we reject H_0. We have convincing evidence that the mean percentage of the time spent playing with the police car is greater for male monkeys than for female monkeys.

b

1. μ_1 = mean percentage of time playing with doll for male monkeys
   μ_2 = mean percentage of time playing with doll for female monkeys
2. H_0: $\mu_1 - \mu_2 = 0$
3. H_a: $\mu_1 - \mu_2 < 0$
4. $\alpha = 0.05$
5. $t = \dfrac{(\bar{x}_1 - \bar{x}_2) - (\text{hypothesized value})}{\sqrt{\dfrac{s_1^2}{n_1} + \dfrac{s_2^2}{n_2}}} = \dfrac{(\bar{x}_1 - \bar{x}_2) - 0}{\sqrt{\dfrac{s_1^2}{n_1} + \dfrac{s_2^2}{n_2}}}$
6. We are told that that it is reasonable to regard these two samples of 44 monkeys as representative of the populations of male and female monkeys. It is therefore reasonable to regard them as random samples. Also $n_1 = 44 \geq 30$ and $n_2 = 44 \geq 30$, so we can proceed with the two-sample t test.
7. $t = \dfrac{9 - 20}{\sqrt{\dfrac{2^2}{44} + \dfrac{4^2}{44}}} = -16.316$
8. $\text{df} = 63.235$
   $P\text{-value} = P(t_{63.235} < -16.316) \approx 0$

9. Since P-value $\approx 0 < 0.05$ we reject H_0. We have convincing evidence that the mean percentage of the time spent playing with the doll is greater for female monkeys than for male monkeys.

c 1. μ_1 = mean percentage of time playing with furry dog for male monkeys
μ_2 = mean percentage of time playing with furry dog for female monkeys
2. H_0: $\mu_1 - \mu_2 = 0$
3. H_a: $\mu_1 - \mu_2 \neq 0$
4. $\alpha = 0.05$
5. $t = \dfrac{(\bar{x}_1 - \bar{x}_2) - (\text{hypothesized value})}{\sqrt{\dfrac{s_1^2}{n_1} + \dfrac{s_2^2}{n_2}}} = \dfrac{(\bar{x}_1 - \bar{x}_2) - 0}{\sqrt{\dfrac{s_1^2}{n_1} + \dfrac{s_2^2}{n_2}}}$
6. We are told that that it is reasonable to regard these two samples of 44 monkeys as representative of the populations of male and female monkeys. It is therefore reasonable to regard them as random samples. Also $n_1 = 44 \geq 30$ and $n_2 = 44 \geq 30$, so we can proceed with the two-sample t test.
7. $t = \dfrac{25 - 20}{\sqrt{\dfrac{5^2}{44} + \dfrac{5^2}{44}}} = 4.690$
8. $\text{df} = 86$
$P\text{-value} = 2 \cdot P(t_{86} > 4.690) \approx 0$
9. Since P-value $\approx 0 < 0.05$ we reject H_0. We have convincing evidence that the mean percentage of the time spent playing with the furry dog is not the same for male monkeys as it is for female monkeys.

d The results do seem to provide convincing evidence of a gender basis in the monkeys' choices of how much time to spend playing with each toy, with the male monkeys spending significantly more time with the "masculine toy" than the female monkeys, and with the female monkeys spending significantly more time with the "feminine toy" than the male monkeys. However, the data also provide convincing evidence of a difference between male and female monkeys in the time they choose to spend playing with a "neutral toy." It is possible that it was some attribute other than masculinity/femininity in the toys that was attracting the different genders of monkey in different ways.

e The given mean time playing with the police car and mean time playing with the doll for female monkeys are sample means for the same sample of female monkeys. The two-sample t test can only be performed when there are two independent random samples.

11.19 **a** Since the samples are small it is necessary to know, or to assume, that the distributions from which the random samples were taken are normal. However, in this case, since both standard deviations are large compared to the means, it seems unlikely that these distributions would have been normal.

b Now, since the samples are large, it is appropriate to carry out the two-sample t test, whatever the distributions from which the samples were taken.

c 1. μ_1 = mean fumonisin level for corn meal made from partially degermed corn

μ_2 = mean fumonisin level for corn meal made from corn that has not been degermed

2. H_0: $\mu_1 - \mu_2 = 0$
3. H_a: $\mu_1 - \mu_2 \neq 0$
4. $\alpha = 0.01$
5. $t = \dfrac{(\bar{x}_1 - \bar{x}_2) - (\text{hypothesized value})}{\sqrt{\dfrac{s_1^2}{n_1} + \dfrac{s_2^2}{n_2}}} = \dfrac{(\bar{x}_1 - \bar{x}_2) - 0}{\sqrt{\dfrac{s_1^2}{n_1} + \dfrac{s_2^2}{n_2}}}$
6. We are told that the samples were random samples from the populations. Also $n_1 = 50 \geq 30$ and $n_2 = 50 \geq 30$, so we can proceed with the two-sample t test.
7. $t = \dfrac{0.59 - 1.21}{\sqrt{\dfrac{1.01^2}{50} + \dfrac{1.71^2}{50}}} = -2.207$
8. $\text{df} = 79.479$
   $P\text{-value} = 2 \cdot P(t_{79.479} < -2.207) = 0.030$
9. Since $P\text{-value} = 0.030 > 0.01$ we do not reject H_0. We do not have convincing evidence that there is a difference in mean fumonisin level for the two types of corn meal.

11.21 **a**

1. μ_1 = mean oxygen consumption for noncourting pairs
   μ_2 = mean oxygen consumption for courting pairs
2. H_0: $\mu_1 - \mu_2 = 0$
3. H_a: $\mu_1 - \mu_2 < 0$
4. $\alpha = 0.05$
5. $t = \dfrac{(\bar{x}_1 - \bar{x}_2) - (\text{hypothesized value})}{\sqrt{\dfrac{s_p^2}{n_1} + \dfrac{s_p^2}{n_2}}} = \dfrac{(\bar{x}_1 - \bar{x}_2) - 0}{\sqrt{\dfrac{s_p^2}{n_1} + \dfrac{s_p^2}{n_2}}}$, where $s_p = \sqrt{\dfrac{(n_1 - 1)s_1^2 + (n_2 - 1)s_2^2}{n_1 + n_2 - 2}}$
6. We need to assume that the samples were random samples from the populations, and that the population distributions are normal. Additionally, the similar sample standard deviations justify our assumption that the populations have equal standard deviations.
7. $s_p = \sqrt{\dfrac{10(0.0066)^2 + 14(0.0071)^2}{24}} = 0.00690$, $t = \dfrac{0.072 - 0.099}{\sqrt{\dfrac{0.00690^2}{11} + \dfrac{0.00690^2}{15}}} = -9.863$
8. $\text{df} = 24$
   $P\text{-value} = P(t_{24} < -9.863) \approx 0$
9. Since $P\text{-value} \approx 0 < 0.05$ we reject H_0. We have convincing evidence that the mean oxygen consumption for courting pairs is higher than the mean oxygen consumption for noncourting pairs.

b For the two-sample t test, $t = -9.979$, $\text{df} = 22.566$, and $P\text{-value} \approx 0$. Thus the conclusion is the same.

11.23 For each pipe, one side (left/right) could be coated with the first type of coating, and the other side could be coated with the other type of coating, with the sides being chosen at random for each pipe. Then the two coatings are being tested under almost exactly equal conditions in terms of the extraneous variables mentioned.

11.25

Swimmer	Water	Guar Syrup	Difference
1	0.9	0.92	−0.02
2	0.92	0.96	−0.04
3	1	0.95	0.05
4	1.1	1.13	−0.03
5	1.2	1.22	−0.02
6	1.25	1.2	0.05
7	1.25	1.26	−0.01
8	1.3	1.3	0
9	1.35	1.34	0.01
10	1.4	1.41	−0.01
11	1.4	1.44	−0.04
12	1.5	1.52	−0.02
13	1.65	1.58	0.07
14	1.7	1.7	0
15	1.75	1.8	−0.05
16	1.8	1.76	0.04
17	1.8	1.84	−0.04
18	1.85	1.89	−0.04
19	1.9	1.88	0.02
20	1.95	1.95	0

1. μ_d = mean swimming velocity difference (water − guar syrup)
2. H_0: $\mu_d = 0$
3. H_a: $\mu_d \neq 0$
4. $\alpha = 0.01$
5. $t = \dfrac{\bar{x}_d - \text{hypothesized value}}{s_d/\sqrt{n}}$
6.

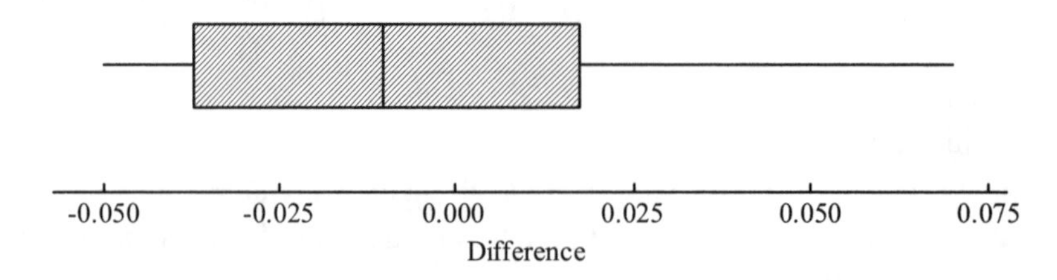

The boxplot shows that the distribution of the differences is roughly symmetrical and has no outliers, so we are justified in assuming that the population distribution of differences is normal. Additionally, we need to assume that this set of differences forms a random sample from the set of differences for all swimmers.

7. $\bar{x}_d = -0.004$, $s_d = 0.035$

$$t = \frac{-0.004 - 0}{0.035/\sqrt{20}} = -0.515$$

8. df = 19
   $P\text{-value} = 2 \cdot P(t_{19} < -0.515) = 0.612$
9. Since $P\text{-value} = 0.612 > 0.01$ we do not reject H_0. We do not have convincing evidence of a difference between the mean swimming speeds in water and guar syrup. The given data are consistent with the authors' conclusion.

11.27

Subject	Location 1 Before	Location 1 After	Difference	Location 2 Before	Location 2 After	Difference
1	6.4	8	-1.6	6.9	9.4	-2.5
2	8.7	12.6	-3.9	9.5	11.2	-1.7
3	7.4	8.4	-1	6.7	10.2	-3.5
4	8.7	9	-0.3	9	9.6	-0.6
5	9.8	8.4	1.4	9.7	9.2	0.5
6	8.9	11	-2.1	9	11.9	-2.9
7	9.3	14.4	-5.1	7.9	9.1	-1.2
8	7.4	11.3	-3.9	8.3	9.3	-1
9	6.6	7.1	-0.5	7.2	8	-0.8
10	8.9	11.2	-2.3	7.4	9.1	-1.7

a
1. μ_d = mean difference in MPF (Location 1, Before − After)
2. H_0: $\mu_d = 0$
3. H_a: $\mu_d < 0$
4. $\alpha = 0.05$
5. $t = \dfrac{\bar{x}_d - \text{hypothesized value}}{s_d/\sqrt{n}}$
6.

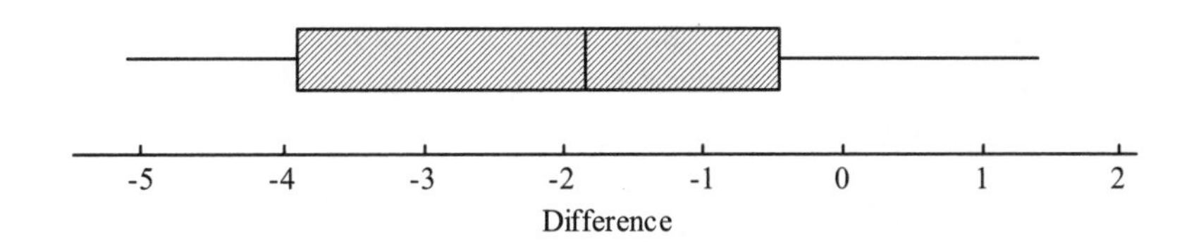

The boxplot shows that the distribution of the differences is roughly symmetrical and has no outliers, so we are justified in assuming that the population distribution of differences is normal. Additionally, we are told to assume that it is reasonable to regard the sample of ten men as representative of healthy adult males, and so we can treat the sample as a random sample from that population.

7. $\bar{x}_d = -1.930,\ s_d = 1.965$
   $$t = \frac{-1.93 - 0}{1.965/\sqrt{10}} = -3.106$$
8. df = 9
   $P\text{-value} = P(t_9 < -3.106) = 0.006$

9. Since $P\text{-value} = 0.006 < 0.05$ we reject H_0. We have convincing evidence that the mean MPF at brain location 1 is higher after diesel exposure.

b Check of Conditions

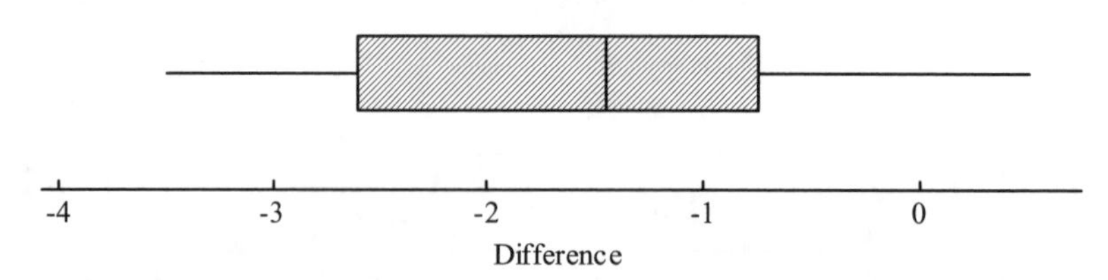

The boxplot shows that the distribution of the differences (before − after at location 2) is roughly symmetrical and has no outliers, so we are justified in assuming that the population distribution of differences is normal. Additionally, we are told to assume that it is reasonable to regard the sample of ten men as representative of healthy adult males, and so we can treat the sample as a random sample from that population.

Calculation

df = 9. The 90% confidence interval for μ_d is

$$\bar{x}_d \pm (t \text{ critical value})\frac{s_d}{\sqrt{n}} = -1.54 \pm 1.833\frac{1.186}{\sqrt{10}} = (\mathbf{-2.228, -0.852})$$

Interpretation

We are 90% confident that the difference in mean MPF at brain location 2 before and after exposure to diesel exhaust is between −2.228 and −0.852.

11.29 a
1. μ_d = mean difference between profile height and actual height (profile − actual)
2. H_0: $\mu_d = 0$
3. H_a: $\mu_d > 0$
4. $\alpha = 0.05$
5. $t = \dfrac{\bar{x}_d - \text{hypothesized value}}{s_d/\sqrt{n}}$
6. We are told to assume that the sample is representative of male online daters, and therefore we are justified in treating it as a random sample. Therefore, since $n = 40 \geq 30$, we can proceed with the paired t test.
7. $t = \dfrac{0.57 - 0}{0.81/\sqrt{40}} = 4.451$
8. df = 39
   $P\text{-value} = P(t_{39} > 4.451) \approx 0$
9. Since $P\text{-value} \approx 0 < 0.05$ we reject H_0. We have convincing evidence that, on average, male online daters overstate their height in online dating profiles.

b Check of Conditions

We are told to assume that the sample is representative of female online daters, and therefore we are justified in treating it as a random sample. Therefore, since $n = 40 \geq 30$, we can proceed with the paired t interval.

Calculation

df = 39. The 95% confidence interval for μ_d is

$$\bar{x}_d \pm (t \text{ critical value})\frac{s_d}{\sqrt{n}} = 0.03 \pm 2.023\frac{0.75}{\sqrt{40}} = (\mathbf{-0.210, 0.270})$$

Interpretation
We are 95% confident that the difference between the mean online dating profile height and mean actual height for female online daters is between −0.210 and 0.270.

c 1. μ_m = mean height difference (profile − actual) for male online daters
μ_f = mean height difference (profile − actual) for female online daters

2. H_0: $\mu_m - \mu_f = 0$

3. H_a: $\mu_m - \mu_f > 0$

4. $\alpha = 0.05$

5. $$t = \frac{(\bar{x}_m - \bar{x}_f) - (\text{hypothesized value})}{\sqrt{\frac{s_m^2}{n_m} + \frac{s_f^2}{n_f}}} = \frac{(\bar{x}_m - \bar{x}_f) - 0}{\sqrt{\frac{s_m^2}{n_m} + \frac{s_f^2}{n_f}}}$$

6. We are told to assume that the samples were representative of the populations, and therefore we are justified in assuming that they are random samples. Also $n_m = 40 \ge 30$ and $n_f = 40 \ge 30$, so we can proceed with the two-sample *t* test.

7. $$t = \frac{0.57 - 0.03}{\sqrt{\frac{0.81^2}{40} + \frac{0.75^2}{40}}} = 3.094$$

8. df = 77.543
$P\text{-value} = P(t_{77.543} > 3.094) = 0.001$

9. Since $P\text{-value} = 0.001 < 0.05$ we reject H_0. We have convincing evidence that $\mu_m - \mu_f > 0$.

d In Part (a), the male profile heights and the male actual heights are paired (according to which individual has the actual height and the height stated in the profile), and with paired samples we use the paired *t* test. In Part (c) we were dealing with two independent samples (the sample of males and the sample of females), and therefore the two-sample *t* test was appropriate.

11.31

Subject	B	P	Difference
1	1928	2126	-198
2	2549	2885	-336
3	2825	2895	-70
4	1924	1942	-18
5	1628	1750	-122
6	2175	2184	-9
7	2114	2164	-50
8	2621	2626	-5
9	1843	2006	-163
10	2541	2627	-86

1. μ_d = mean bone mineral content difference (breast feeding − postweaning)
2. H_0: $\mu_d = -25$
3. H_a: $\mu_d < -25$
4. $\alpha = 0.05$
5. $t = \dfrac{\bar{x}_d - \text{hypothesized value}}{s_d/\sqrt{n}}$
6.

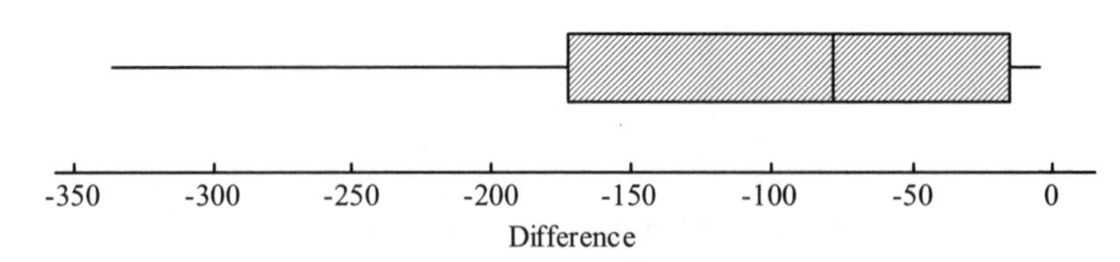

The boxplot shows that the distribution of the sample differences is negatively skewed, but for a relatively small sample this distribution is not inconsistent with a population that is normally distributed. Additionally, the sample distribution of differences has no outliers. We need to assume that the mothers used in the study formed a random sample from the population of mothers.

7. $\bar{x}_d = -105.7$, $s_d = 103.845$

   $t = \dfrac{-105.7-(-25)}{103.845/\sqrt{10}} = -2.457$
8. df = 9

   $P\text{-value} = P(t_9 < -2.457) = 0.018$
9. Since $P\text{-value} = 0.018 < 0.05$ we reject H_0. We have convincing evidence that the average total body bone mineral content during postweaning is greater than that during breast feeding by more than 25 grams.

11.33 **a**

1. μ_d = mean difference in wrist extension (type A − type B)
2. H_0: $\mu_d = 0$
3. H_a: $\mu_d > 0$
4. $\alpha = 0.05$
5. $t = \dfrac{\bar{x}_d - \text{hypothesized value}}{s_d/\sqrt{n}}$
6. We are told to assume that the sample is representative of the population of computer users, and therefore we are justified in treating it as a random sample from that population. However, in order to proceed with the paired *t* test we need to assume that the population of differences is normally distributed.
7. $t = \dfrac{8.82-0}{10/\sqrt{24}} = 4.321$
8. df = 23

   $P\text{-value} = P(t_{23} > 4.321) \approx 0$
9. Since $P\text{-value} \approx 0 < 0.05$ we reject H_0. We have convincing evidence that the mean wrist extension for mouse type A is greater than for mouse type B.

b Now $t = \dfrac{8.82 - 0}{26/\sqrt{24}} = 1.662$, and so $P\text{-value} = P(t_{23} > 1.662) = 0.055$. Since $P\text{-value} = 0.055 > 0.05$ we do not reject H_0. We do not have convincing evidence that the mean wrist extension for mouse type A is greater than for mouse type B.

c A lower standard deviation in the sample of differences means that we have a lower estimate of the standard deviation of the population of differences. Assuming that the mean wrist extensions for the two mouse types are the same (in other words, that the mean of the population of differences is zero), a sample mean difference of as much as 8.82 is much less likely when the standard deviation of the population of differences is around 10 than when the standard deviation of the population of differences is around 26.

11.35 μ_d = mean difference between verbal ability score at age 8 and verbal ability score at age 3 (age 8 − age 3)
H_0: $\mu_d = 0$
H_a: $\mu_d > 0$
$\alpha = 0.05$
We are told to assume that the sample is a random sample from the population of children born prematurely. Therefore, since $n = 50 \geq 30$, we can proceed with the paired *t* test.
$P\text{-value} = 0.001$
Since $P\text{-value} = 0.001 < 0.05$ we reject H_0. We have convincing evidence that the mean verbal ability score for children born prematurely increases between age 3 and age 8.

11.37
1. p_1 = proportion of guests who reserve by phone who are satisfied
   p_2 = proportion of guests who reserve online who are satisfied
2. H_0: $p_1 - p_2 = 0$
3. H_a: $p_1 - p_2 < 0$
4. $\alpha = 0.05$
5. $$z = \frac{\hat{p}_1 - \hat{p}_2}{\sqrt{\dfrac{\hat{p}_c(1-\hat{p}_c)}{n_1} + \dfrac{\hat{p}_c(1-\hat{p}_c)}{n_2}}}$$
6. We are told that the samples were independent random samples from the populations. Also $n_1\hat{p}_1 = 80(57/80) = 57 \geq 10$, $n_1(1-\hat{p}_1) = 80(23/80) = 23 \geq 10$, $n_2\hat{p}_2 = 60(50/60) = 50 \geq 10$, and $n_2(1-\hat{p}_2) = 60(10/60) = 10 \geq 10$, so the samples are large enough.
7. $\hat{p}_c = \dfrac{57+50}{80+60} = \dfrac{107}{140}$
   $$z = \frac{57/80 - 50/60}{\sqrt{\dfrac{(107/140)(33/140)}{80} + \dfrac{(107/140)(33/140)}{60}}} = -1.667$$
8. $P\text{-value} = P(Z < -1.667) = 0.048$
9. Since $P\text{-value} = 0.048 < 0.05$ we reject H_0. We have convincing evidence that the proportion who are satisfied is higher for those who reserve online than for those who reserve by phone.

11.39 **a**
1. p_1 = proportion of Gen Y respondents who donated by text message
   p_2 = proportion of Gen X respondents who donated by text message

2. H_0: $p_1 - p_2 = 0$
3. H_a: $p_1 - p_2 > 0$
4. $\alpha = 0.01$
5. $z = \dfrac{\hat{p}_1 - \hat{p}_2}{\sqrt{\dfrac{\hat{p}_c(1-\hat{p}_c)}{n_1} + \dfrac{\hat{p}_c(1-\hat{p}_c)}{n_2}}}$
6. We are told to regard the samples as representative of the Gen Y and Gen X populations, so it is reasonable to treat them as independent random samples from the populations. Also $n_1\hat{p}_1 = 400(0.17) = 68 \ge 10$, $n_1(1-\hat{p}_1) = 400(0.83) = 332 \ge 10$, $n_2\hat{p}_2 = 400(0.14) = 56 \ge 10$, and $n_2(1-\hat{p}_2) = 400(0.86) = 344 \ge 10$, so the samples are large enough.
7. $\hat{p}_c = \dfrac{n_1\hat{p}_1 + n_2\hat{p}_2}{n_1 + n_2} = \dfrac{400(0.17) + 400(0.14)}{400 + 400} = 0.155$

   $$z = \frac{0.17 - 0.14}{\sqrt{\dfrac{(0.155)(0.845)}{400} + \dfrac{(0.155)(0.845)}{400}}} = 1.172$$
8. $P\text{-value} = P(Z > 1.172) = 0.121$
9. Since $P\text{-value} = 0.121 > 0.01$ we do not reject H_0. We do not have convincing evidence that the proportion of those in Gen Y who donated to Haiti relief via text message is greater than the proportion of those in Gen X.

b Check of Conditions

See Part (a).

Calculation

The 99% confidence interval for $p_1 - p_2$ is

$$(\hat{p}_1 - \hat{p}_2) \pm (z \text{ critical value})\sqrt{\frac{\hat{p}_1(1-\hat{p}_1)}{n_1} + \frac{\hat{p}_2(1-\hat{p}_2)}{n_2}}$$

$$= (0.17 - 0.14) \pm 2.576\sqrt{\frac{(0.17)(0.83)}{400} + \frac{(0.14)(0.86)}{400}}$$

$$= (\mathbf{-0.036, 0.096})$$

Interpretation of Interval

We are 99% confident that the difference between the proportion of Gen Y and the proportion of Gen X who made a donation via text message is between −0.036 and 0.096.

Interpretation of Confidence Level

In repeated sampling with random samples of size 400, 99% of the resulting confidence intervals would contain the true difference in proportions who donated via text message.

11.41 a

1. p_1 = proportion of American teenage girls who say that newspapers are boring

   p_2 = proportion of American teenage boys who say that newspapers are boring
2. H_0: $p_1 - p_2 = 0$
3. H_a: $p_1 - p_2 \ne 0$
4. $\alpha = 0.05$

5. $z = \dfrac{\hat{p}_1 - \hat{p}_2}{\sqrt{\dfrac{\hat{p}_c(1-\hat{p}_c)}{n_1} + \dfrac{\hat{p}_c(1-\hat{p}_c)}{n_2}}}$

6. The samples were representative of the populations, so it is reasonable to treat them as independent random samples from the populations. Also $n_1\hat{p}_1 = 58(0.41) = 24 \geq 10$, $n_1(1-\hat{p}_1) = 58(0.59) = 34 \geq 10$, $n_2\hat{p}_2 = 41(0.44) = 18 \geq 10$, and $n_2(1-\hat{p}_2) = 41(0.56) = 23 \geq 10$, so the samples are large enough.

7. $\hat{p}_c = \dfrac{n_1\hat{p}_1 + n_2\hat{p}_2}{n_1 + n_2} = \dfrac{58(0.41)+41(0.44)}{58+41} = 0.422$

$$z = \frac{0.41 - 0.44}{\sqrt{\dfrac{(0.422)(0.578)}{58} + \dfrac{(0.422)(0.578)}{41}}} = -0.298$$

8. $P\text{-value} = 2 \cdot P(Z < -0.298) = 0.766$

9. Since $P\text{-value} = 0.766 > 0.05$ we do not reject H_0. We do not have convincing evidence that the proportion of girls who say that newspapers are boring is different from the proportion of boys who say that newspapers are boring.

b Since the samples are larger than in Part (a), the conditions for performing the test are also satisfied here. The calculations will change to the following:

$$\hat{p}_c = \frac{n_1\hat{p}_1 + n_2\hat{p}_2}{n_1 + n_2} = \frac{2000(0.41)+2500(0.44)}{2000+2500} = 0.427$$

$$z = \frac{0.41 - 0.44}{\sqrt{\dfrac{(0.427)(0.573)}{2000} + \dfrac{(0.427)(0.573)}{2500}}} = -2.022$$

$P\text{-value} = 2 \cdot P(Z < -2.022) = 0.043$

Since $P\text{-value} = 0.043 < 0.05$ we reject H_0. We have convincing evidence that the proportion of girls who say that newspapers are boring is different from the proportion of boys who say that newspapers are boring.

c Assuming that the population proportions are equal, you are much less likely to get a difference in sample proportions as large as the one given when the samples are very large than when the samples are relatively small, since large samples are likely to give more accurate estimates of the population proportions. Therefore, when the given difference in sample proportions was based on larger samples, this produced stronger evidence of a difference in population proportions.

11.43 a Check of Conditions

We are told to regard the samples as representative of teens before and after the ban, so it is reasonable to treat them as independent random samples from these populations. Also $n_1\hat{p}_1 = 200(0.11) = 22 \geq 10$, $n_1(1-\hat{p}_1) = 200(0.89) = 178 \geq 10$, $n_2\hat{p}_2 = 150(0.12) = 18 \geq 10$, and $n_2(1-\hat{p}_2) = 150(0.88) = 132 \geq 10$, so the samples are large enough.

Calculation

The 95% confidence interval for $p_1 - p_2$ is

$$(\hat{p}_1-\hat{p}_2)\pm(z\text{ critical value})\sqrt{\frac{\hat{p}_1(1-\hat{p}_1)}{n_1}+\frac{\hat{p}_2(1-\hat{p}_2)}{n_2}}$$

$$=(0.11-0.12)\pm1.96\sqrt{\frac{(0.11)(0.89)}{200}+\frac{(0.12)(0.88)}{150}}$$

$$=(\mathbf{-0.078,0.058})$$

Interpretation

We are 95% confident that the difference between the proportion of teenagers using a cell phone before the ban and the proportion of teenagers using a cell phone after the ban is between −0.078 and 0.058.

b Zero is included in the confidence interval. This tell us that we do not have convincing evidence at a 0.05 significance level of a difference between the proportion of teenagers using a cell phone before the ban and the proportion of teenagers using a cell phone after the ban.

11.45 No. It is not appropriate to use the two-sample z test because the groups are not large enough. We are not told the sizes of the groups, but we know that each is at most 81. The sample proportion for the fish oil group is 0.05, and 81(0.05) = 4.05, which is less than 10. So the conditions for the two-sample z test are not satisfied.

11.47

1. p_1 = proportion of passengers on airplanes that do not recirculate air who have post-flight respiratory symptoms
   p_2 = proportion of passengers on airplanes that recirculate air who have post-flight respiratory symptoms
2. H_0: $p_1-p_2=0$
3. H_a: $p_1-p_2\neq0$
4. $\alpha=0.05$
5. $$z=\frac{\hat{p}_1-\hat{p}_2}{\sqrt{\frac{\hat{p}_c(1-\hat{p}_c)}{n_1}+\frac{\hat{p}_c(1-\hat{p}_c)}{n_2}}}$$
6. We are told to assume that it is reasonable to regard the two samples as being independently selected and as representative of the two populations. Therefore it is reasonable to treat the samples as independent random samples from the populations. Also, $n_1\hat{p}_1=517(108/517)=108\geq10$, $n_1(1-\hat{p}_1)=517(409/517)=409\geq10$, $n_2\hat{p}_2=583(111/583)=111\geq10$, and $n_2(1-\hat{p}_2)=583(472/583)=472\geq10$, so the samples are large enough.
7. $$\hat{p}_c=\frac{108+111}{517+583}=\frac{219}{1100}$$
   $$z=\frac{108/517-111/583}{\sqrt{\frac{(219/1100)(881/1100)}{517}+\frac{(219/1100)(881/1100)}{583}}}=0.767$$
8. $P\text{-value}=2\cdot P(Z>0.767)=0.443$
9. Since $P\text{-value}=0.443>0.05$ we do not reject H_0. We do not have convincing evidence that the proportion of passengers with post-flight respiratory symptoms differs for planes that do and do not recirculate air.

11.49 Check of Conditions

We are told that the samples were random samples from the populations of Americans age 12 and over. Also, $n_1\hat{p}_1 = 1112(0.2) = 222 \geq 10$, $n_1(1-\hat{p}_1) = 1112(0.8) = 890 \geq 10$, $n_2\hat{p}_2 = 1112(0.15) = 167 \geq 10$, and $n_2(1-\hat{p}_2) = 1112(0.85) = 945 \geq 10$, so the samples are large enough.

Calculation

The 95% confidence interval for $p_1 - p_2$ is

$$(\hat{p}_1 - \hat{p}_2) \pm (z \text{ critical value})\sqrt{\frac{\hat{p}_1(1-\hat{p}_1)}{n_1} + \frac{\hat{p}_2(1-\hat{p}_2)}{n_2}}$$

$$= (0.2 - 0.15) \pm 1.96\sqrt{\frac{(0.2)(0.8)}{1112} + \frac{(0.15)(0.85)}{1112}}$$

$$= (\mathbf{0.018, 0.082})$$

Interpretation

We are 95% confident that the proportion of Americans age 12 and over who owned an MP3 player in 2006 minus the proportion of Americans age 12 and over who owned an MP3 player in 2005 is between 0.018 and 0.082.

Zero is not included in the confidence interval. This means that we have convincing evidence at the 0.05 significance level of a difference between the proportions of people owning MP3 players in 2006 and 2005.

11.51

1. p_1 = proportion of parents who think that science and higher math are essential
   p_2 = proportion of students in grades 6–12 who think that science and higher math are essential
2. H_0: $p_1 - p_2 = 0$
3. H_a: $p_1 - p_2 \neq 0$
4. $\alpha = 0.05$
5. $$z = \frac{\hat{p}_1 - \hat{p}_2}{\sqrt{\frac{\hat{p}_c(1-\hat{p}_c)}{n_1} + \frac{\hat{p}_c(1-\hat{p}_c)}{n_2}}}$$
6. We are told that the samples were independently selected, but we need to assume that they were independent *random* samples from the populations. Also $n_1\hat{p}_1 = 1379(0.62) = 855 \geq 10$, $n_1(1-\hat{p}_1) = 1379(0.38) = 524 \geq 10$, $n_2\hat{p}_2 = 1342(0.5) = 671 \geq 10$, and $n_2(1-\hat{p}_2) = 1342(0.5) = 671 \geq 10$, so the samples are large enough.
7. $$\hat{p}_c = \frac{n_1\hat{p}_1 + n_2\hat{p}_2}{n_1 + n_2} = \frac{1379(0.62) + 1342(0.5)}{1379 + 1342} = 0.561$$
   $$z = \frac{0.62 - 0.5}{\sqrt{\frac{(0.561)(0.439)}{1379} + \frac{(0.561)(0.439)}{1342}}} = 6.306$$
8. $P\text{-value} = 2 \cdot P(Z > 6.306) \approx 0$
9. Since $P\text{-value} \approx 0 < 0.05$ we reject H_0. We have convincing evidence that the proportion of parents who regard science and mathematics as crucial is different from the corresponding proportion of students in grades 6–12.

11.53 1. p_1 = proportion of college graduates who have sunburn
p_2 = proportion of people without a high school degree who have sunburn

2. H_0: $p_1 - p_2 = 0$
3. H_a: $p_1 - p_2 > 0$
4. $\alpha = 0.05$
5. $$z = \frac{\hat{p}_1 - \hat{p}_2}{\sqrt{\frac{\hat{p}_c(1-\hat{p}_c)}{n_1} + \frac{\hat{p}_c(1-\hat{p}_c)}{n_2}}}$$
6. We are told to assume that the samples were random samples from the populations. Also $n_1\hat{p}_1 = 200(0.43) = 86 \geq 10$, $n_1(1-\hat{p}_1) = 200(0.57) = 114 \geq 10$, $n_2\hat{p}_2 = 200(0.25) = 50 \geq 10$, and $n_2(1-\hat{p}_2) = 200(0.75) = 150 \geq 10$, so the samples are large enough.
7. $$\hat{p}_c = \frac{n_1\hat{p}_1 + n_2\hat{p}_2}{n_1 + n_2} = \frac{200(0.43) + 200(0.25)}{200 + 200} = 0.34$$
$$z = \frac{0.43 - 0.25}{\sqrt{\frac{(0.34)(0.66)}{200} + \frac{(0.34)(0.66)}{200}}} = 3.800$$
8. $P\text{-value} = P(Z > 3.800) \approx 0$
9. Since $P\text{-value} \approx 0 < 0.05$ we reject H_0. We have convincing evidence that the proportion experiencing sunburn is greater for college graduates than it is for those without a high school degree.

11.55 **a** 1. p_1 = proportion of Austrian avid mountain bikers who have a low sperm count
p_2 = proportion of Austrian nonbikers who have a low sperm count

2. H_0: $p_1 - p_2 = 0$
3. H_a: $p_1 - p_2 > 0$
4. $\alpha = 0.05$
5. $$z = \frac{\hat{p}_1 - \hat{p}_2}{\sqrt{\frac{\hat{p}_c(1-\hat{p}_c)}{n_1} + \frac{\hat{p}_c(1-\hat{p}_c)}{n_2}}}$$
6. We are told to assume that the percentages were based on independent samples and that the samples were representative of Austrian avid mountain bikers and nonbikers. So it is reasonable to assume that the samples were independent *random* samples. Also $n_1\hat{p}_1 = 100(0.9) = 90 \geq 10$, $n_1(1-\hat{p}_1) = 100(0.1) = 10 \geq 10$, $n_2\hat{p}_2 = 100(0.26) = 26 \geq 10$, and $n_2(1-\hat{p}_2) = 100(0.74) = 74 \geq 10$, so the samples are large enough.
7. $$\hat{p}_c = \frac{n_1\hat{p}_1 + n_2\hat{p}_2}{n_1 + n_2} = \frac{100(0.9) + 100(0.26)}{100 + 100} = 0.58$$
$$z = \frac{0.9 - 0.26}{\sqrt{\frac{(0.58)(0.42)}{100} + \frac{(0.58)(0.42)}{100}}} = 9.169$$
8. $P\text{-value} = P(Z > 9.169) \approx 0$
9. Since $P\text{-value} \approx 0 < 0.05$ we reject H_0. We have convincing evidence that the proportion of Austrian avid mountain bikers with a low sperm count is higher than the equivalent proportion for Austrian nonbikers.

b No. Since this is an observational study, causation cannot be inferred from the result. It could be suggested that, for example, Austrian men who have low sperm counts have a tendency to choose mountain biking as a hobby.

11.57 Since the data given are population characteristics an inference procedure is not applicable. It is *known* that the rate of Lou Gehrig's disease amongst soldiers sent to the war is higher than for those not sent to the war.

11.59 **a** First hypothesis test

p_1 = proportion of those receiving the intervention whose references to sex decrease to zero

p_2 = proportion of those not receiving the intervention whose references to sex decrease to zero

H_0: $p_1 - p_2 = 0$

H_a: $p_1 - p_2 \neq 0$

(Note: We know that the researchers were using two-sided alternative hypotheses, otherwise the P-value greater than 0.5 in the second hypothesis test would not have been possible for the given results.)

Since P-value = 0.05, H_0 is rejected at the 0.05 level.

Second hypothesis test

p_1 = proportion of those receiving the intervention whose references to substance abuse decrease to zero

p_2 = proportion of those not receiving the intervention whose references to substance abuse decrease to zero

H_0: $p_1 - p_2 = 0$

H_a: $p_1 - p_2 \neq 0$

Since P-value = 0.61, H_0 is not rejected at the 0.05 level.

Third hypothesis test

p_1 = proportion of those receiving the intervention whose profiles are set to "private" at follow-up

p_2 = proportion of those not receiving the intervention whose profiles are set to "private" at follow-up

H_0: $p_1 - p_2 = 0$

H_a: $p_1 - p_2 \neq 0$

Since P-value = 0.45, H_0 is not rejected at the 0.05 level.

Fourth hypothesis test

p_1 = proportion of those receiving the intervention whose profiles show any of the three protective changes

p_2 = proportion of those not receiving the intervention whose profiles show any of the three protective changes

H_0: $p_1 - p_2 = 0$

H_a: $p_1 - p_2 \neq 0$

Since P-value = 0.07, H_0 is not rejected at the 0.05 level.

b If we want to know whether the email intervention *reduces* (as opposed to *changes*) adolescents' display of risk behavior in their profiles, then we use one-sided alternative hypotheses and the *P*-values are halved. If that is the case, using a 0.05 significance level, we are convinced that the intervention is effective with regard to reduction of references to sex and that the proportion showing any of the three protective changes is greater for those receiving the email intervention. Each of the other two apparently reduced proportions could have occurred by chance.

11.61 **a**

1. μ_1 = mean appropriateness score assigned to wearing a hat in class for students
   μ_2 = mean appropriateness score assigned to wearing a hat in class for faculty
2. H_0: $\mu_1 - \mu_2 = 0$
3. H_a: $\mu_1 - \mu_2 \neq 0$
4. $\alpha = 0.05$
5. $t = \dfrac{(\bar{x}_1 - \bar{x}_2) - (\text{hypothesized value})}{\sqrt{\dfrac{s_1^2}{n_1} + \dfrac{s_2^2}{n_2}}} = \dfrac{(\bar{x}_1 - \bar{x}_2) - 0}{\sqrt{\dfrac{s_1^2}{n_1} + \dfrac{s_2^2}{n_2}}}$
6. We are told that the samples were random samples from the populations. Also $n_1 = 173 \geq 30$ and $n_2 = 98 \geq 30$, so we can proceed with the two-sample *t* test.
7. $t = \dfrac{2.80 - 3.63}{\sqrt{\dfrac{1.0^2}{173} + \dfrac{1.0^2}{98}}} = -6.565$
8. $\text{df} = 201.549$
   $P\text{-value} = 2 \cdot P(t_{201.549} < -6.565) \approx 0$
9. Since $P\text{-value} \approx 0 < 0.05$ we reject H_0. We have convincing evidence that the mean appropriateness score assigned to wearing a hat in class differs for students and faculty.

b

1. μ_1 = mean appropriateness score assigned to addressing an instructor by his/her first name for students
   μ_2 = mean appropriateness score assigned to addressing an instructor by his/her first name for faculty
2. H_0: $\mu_1 - \mu_2 = 0$
3. H_a: $\mu_1 - \mu_2 > 0$
4. $\alpha = 0.05$
5. $t = \dfrac{(\bar{x}_1 - \bar{x}_2) - (\text{hypothesized value})}{\sqrt{\dfrac{s_1^2}{n_1} + \dfrac{s_2^2}{n_2}}} = \dfrac{(\bar{x}_1 - \bar{x}_2) - 0}{\sqrt{\dfrac{s_1^2}{n_1} + \dfrac{s_2^2}{n_2}}}$
6. We are told that the samples were random samples from the populations. Also $n_1 = 173 \geq 30$ and $n_2 = 98 \geq 30$, so we can proceed with the two-sample *t* test.
7. $t = \dfrac{2.90 - 2.11}{\sqrt{\dfrac{1.0^2}{173} + \dfrac{1.0^2}{98}}} = 6.249$
8. $\text{df} = 201.549$
   $P\text{-value} = P(t_{201.549} > 6.249) \approx 0$

9. Since $P\text{-value} \approx 0 < 0.05$ we reject H_0. We have convincing evidence that the mean appropriateness score assigned to addressing an instructor by his/her first name is greater for students than for faculty.

c

1. μ_1 = mean appropriateness score assigned to talking on a cell phone in class for students
   μ_2 = mean appropriateness score assigned to talking on a cell phone in class for faculty
2. H_0: $\mu_1 - \mu_2 = 0$
3. H_a: $\mu_1 - \mu_2 \neq 0$
4. $\alpha = 0.05$
5. $t = \dfrac{(\bar{x}_1 - \bar{x}_2) - (\text{hypothesized value})}{\sqrt{\dfrac{s_1^2}{n_1} + \dfrac{s_2^2}{n_2}}} = \dfrac{(\bar{x}_1 - \bar{x}_2) - 0}{\sqrt{\dfrac{s_1^2}{n_1} + \dfrac{s_2^2}{n_2}}}$
6. We are told that the samples were random samples from the populations. Also $n_1 = 173 \geq 30$ and $n_2 = 98 \geq 30$, so we can proceed with the two-sample t test.
7. $t = \dfrac{1.11 - 1.10}{\sqrt{\dfrac{1.0^2}{173} + \dfrac{1.0^2}{98}}} = 0.079$
8. $\text{df} = 201.549$
   $P\text{-value} = 2 \cdot P(t_{201.549} > 0.079) = 0.937$
9. Since $P\text{-value} = 0.469 > 0.05$ we do not reject H_0. We do not have convincing evidence that the mean appropriateness score assigned to talking on a cell phone in class differs for students and faculty.

No, this does not imply that students and faculty consider it acceptable to talk on a cell phone during class, in fact the low sample mean ratings for both students and faculty show that both groups on the whole feel that the behavior is inappropriate.

11.63

Initial		After 9 Days			
Treatment	**Control**	**Treatment**	**Control**	**Difference (Init – 9 Day, for Trtmnt Grp)**	**Difference (Init – 9 Day, for Cntrl Grp)**
11.4	9.1	138.3	9.3	-126.9	-0.2
9.6	8.7	104	8.8	-94.4	-0.1
10.1	9.7	96.4	8.8	-86.3	0.9
8.5	10.8	89	10.1	-80.5	0.7
10.3	10.9	88	9.6	-77.7	1.3
10.6	10.6	103.8	8.6	-93.2	2
11.8	10.1	147.3	10.4	-135.5	-0.3
9.8	12.3	97.1	12.4	-87.3	-0.1
10.9	8.8	172.6	9.3	-161.7	-0.5
10.3	10.4	146.3	9.5	-136	0.9
10.2	10.9	99	8.4	-88.8	2.5
11.4	10.4	122.3	8.7	-110.9	1.7
9.2	11.6	103	12.5	-93.8	-0.9
10.6	10.9	117.8	9.1	-107.2	1.8
10.8		121.5		-110.7	
8.2		93		-84.8	

a

1. μ_d = mean difference in selenium level (initial level – 9-day level) for cows receiving supplement
2. H_0: $\mu_d = 0$
3. H_a: $\mu_d < 0$
4. $\alpha = 0.05$
5. $t = \dfrac{\bar{x}_d - \text{hypothesized value}}{s_d/\sqrt{n}}$
6.

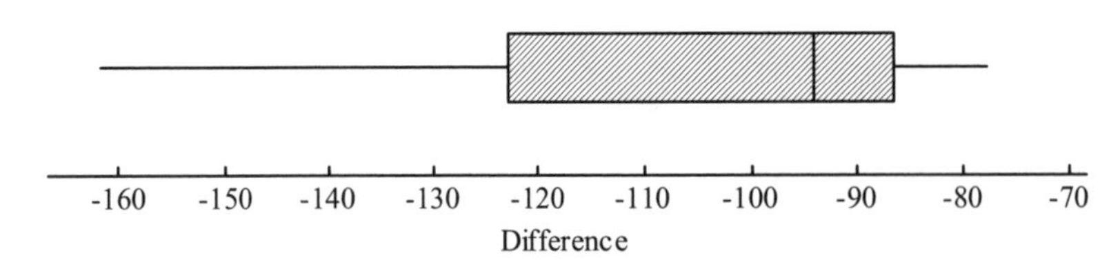

The boxplot shows a distribution of sample differences that is negatively skewed, but in a small sample (along with the fact that there are no outliers) this is nonetheless consistent with an assumption of normality in the population of differences. Additionally, we need to assume that the cows who received the supplement form a random sample from the set of all cows.

7. $\bar{x}_d = -104.731$, $s_d = 24.101$

$$t = \frac{-104.731 - 0}{24.101/\sqrt{16}} = -17.382$$

8. df = 15

$P\text{-value} = P(t_{15} < -17.382) \approx 0$

9. Since P-value $\approx 0 < 0.05$ we reject H_0. We have convincing evidence that the mean selenium concentration is greater after 9 days of the selenium supplement.

b 1. μ_d = mean difference in selenium level (initial level − 9-day level) for cows not receiving supplement
2. H_0: $\mu_d = 0$
3. H_a: $\mu_d \neq 0$
4. $\alpha = 0.05$
5. $t = \dfrac{\bar{x}_d - \text{hypothesized value}}{s_d/\sqrt{n}}$
6.

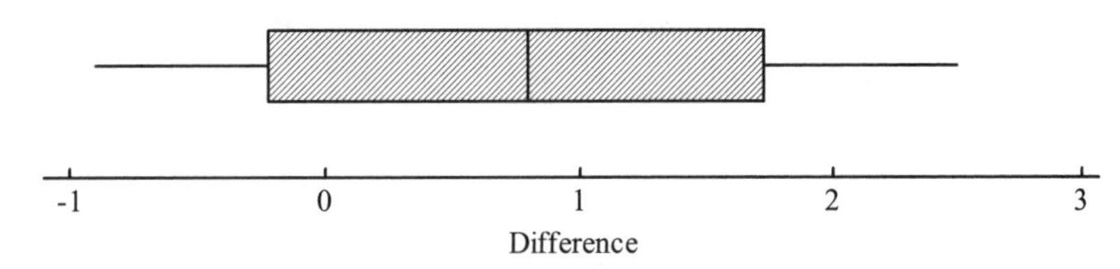

Since the boxplot is roughly symmetrical and there are no outliers we are justified in assuming a normal distribution for the population of differences. Additionally, we need to assume that the cows who did not receive the supplement form a random sample from the set of all cows.

7. $\bar{x}_d = 0.693,\ s_d = 1.062$

$t = \dfrac{0.693 - 0}{1.062/\sqrt{14}} = 2.440$

8. df = 13

$P\text{-value} = 2 \cdot P(t_{13} > 2.440) = 0.030$

9. Since P-value $= 0.030 < 0.05$ we reject H_0. At the 0.05 level the results are inconsistent with the hypothesis of no significant change in mean selenium concentration over the 9-day period for cows that did not receive the supplement.

c No, the paired t test would not be appropriate since the treatment and control groups were not paired samples.

11.65 1. p_1 = proportion of resumes with "white-sounding" names that receive responses

p_2 = proportion of resumes with "black-sounding" names that receive responses

2. H_0: $p_1 - p_2 = 0$
3. H_a: $p_1 - p_2 > 0$
4. $\alpha = 0.05$
5. $z = \dfrac{\hat{p}_1 - \hat{p}_2}{\sqrt{\dfrac{\hat{p}_c(1-\hat{p}_c)}{n_1} + \dfrac{\hat{p}_c(1-\hat{p}_c)}{n_2}}}$
6. We need to assume that the 5000 jobs applied for were randomly assigned to the names used. Also, $n_1\hat{p}_1 = 2500(250/2500) = 250 \geq 10$, $n_1(1-\hat{p}_1) = 2500(2250/2500) = 2250 \geq 10$,

$n_2\hat{p}_2 = 2500(167/2500) = 167 \geq 10$, and $n_2(1-\hat{p}_2) = 2500(2333/2500) = 2333 \geq 10$, so the samples are large enough.

7. $\hat{p}_c = \dfrac{250+167}{2500+2500} = \dfrac{417}{5000}$

$$z = \frac{250/2500 - 167/2500}{\sqrt{\dfrac{(417/5000)(4583/5000)}{2500} + \dfrac{(417/5000)(4583/5000)}{2500}}} = 4.245$$

8. $P\text{-value} = P(Z > 4.245) \approx 0$
9. Since $P\text{-value} \approx 0 < 0.05$ we reject H_0. We have convincing evidence that the proportion eliciting responses is higher for "white-sounding" first names.

11.67 a

1. μ_1 = mean elongation for a square knot for Maxon thread
   μ_2 = mean elongation for a Duncan loop for Maxon thread
2. H_0: $\mu_1 - \mu_2 = 0$
3. H_a: $\mu_1 - \mu_2 \neq 0$
4. $\alpha = 0.05$
5. $t = \dfrac{(\bar{x}_1 - \bar{x}_2) - (\text{hypothesized value})}{\sqrt{\dfrac{s_1^2}{n_1} + \dfrac{s_2^2}{n_2}}} = \dfrac{(\bar{x}_1 - \bar{x}_2) - 0}{\sqrt{\dfrac{s_1^2}{n_1} + \dfrac{s_2^2}{n_2}}}$
6. We are told that the types of knot and suture material were randomly assigned to the specimens. We are also told to assume that the relevant elongation distributions are approximately normal.
7. $t = \dfrac{10-11}{\sqrt{\dfrac{0.1^2}{10} + \dfrac{0.3^2}{15}}} = -11.952$
8. df = 18.266
   $P\text{-value} = 2 \cdot P(t_{18.266} < -11.952) \approx 0$
9. Since $P\text{-value} \approx 0 < 0.05$ we reject H_0. We have convincing evidence that the mean elongations for the square knot and the Duncan loop for Maxon thread are different.

b

1. μ_1 = mean elongation for a square knot for Ticron thread
   μ_2 = mean elongation for a Duncan loop for Ticron thread
2. H_0: $\mu_1 - \mu_2 = 0$
3. H_a: $\mu_1 - \mu_2 \neq 0$
4. $\alpha = 0.05$
5. $t = \dfrac{(\bar{x}_1 - \bar{x}_2) - (\text{hypothesized value})}{\sqrt{\dfrac{s_1^2}{n_1} + \dfrac{s_2^2}{n_2}}} = \dfrac{(\bar{x}_1 - \bar{x}_2) - 0}{\sqrt{\dfrac{s_1^2}{n_1} + \dfrac{s_2^2}{n_2}}}$
6. We are told that the types of knot and suture material were randomly assigned to the specimens. We are also told to assume that the relevant elongation distributions are approximately normal.

7. $t = \dfrac{2.5-10.9}{\sqrt{\dfrac{0.06^2}{10}+\dfrac{0.4^2}{11}}} = -68.803$

8. df = 10.494

   P-value $= 2 \cdot P(t_{10.494} < -68.803) \approx 0$

9. Since P-value $\approx 0 < 0.05$ we reject H_0. We have convincing evidence that the mean elongations for the square knot and the Duncan loop for Ticron thread are different.

c 1. μ_1 = mean elongation for a Duncan loop for Maxon thread

   μ_2 = mean elongation for a Duncan loop for Ticron thread

2. H_0: $\mu_1 - \mu_2 = 0$

3. H_a: $\mu_1 - \mu_2 \neq 0$

4. $\alpha = 0.05$

5. $t = \dfrac{(\bar{x}_1 - \bar{x}_2) - (\text{hypothesized value})}{\sqrt{\dfrac{s_1^2}{n_1}+\dfrac{s_2^2}{n_2}}} = \dfrac{(\bar{x}_1 - \bar{x}_2) - 0}{\sqrt{\dfrac{s_1^2}{n_1}+\dfrac{s_2^2}{n_2}}}$

6. We are told that the types of knot and suture material were randomly assigned to the specimens. We are also told to assume that the relevant elongation distributions are approximately normal.

7. $t = \dfrac{11-10.9}{\sqrt{\dfrac{0.3^2}{15}+\dfrac{0.4^2}{11}}} = 0.698$

8. df = 17.789

   P-value $= 2 \cdot P(t_{17.789} > 0.698) = 0.494$

9. Since P-value $= 0.697 > 0.05$ we do not reject H_0. We do not have convincing evidence that the mean elongations for the Duncan loop for Maxon thread and Ticron thread are different.

11.69

Subject	1	2	3	4	5	6	7	8
1 hr later	14	12	18	7	11	9	16	15
24 hr later	10	4	14	6	9	6	12	12
Difference	4	8	4	1	2	3	4	3

1. μ_d = mean difference in number of objects remembered (1 hr − 24 hours)
2. H_0: $\mu_d = 3$
3. H_a: $\mu_d > 3$
4. $\alpha = 0.01$
5. $t = \dfrac{\bar{x}_d - \text{hypothesized value}}{s_d / \sqrt{n}}$
6.

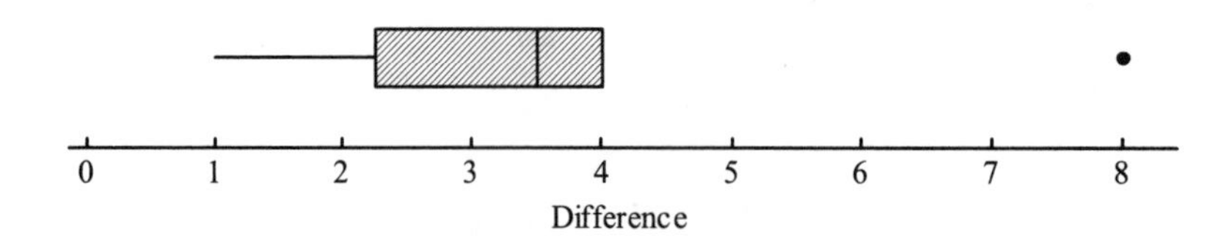

The boxplot is roughly symmetrical but there is one outlier. Nonetheless we will assume that the population distribution of differences is normal. We are told that the eight students were selected at random from the large psychology class.

7. $\bar{x}_d = 3.625,\ s_d = 2.066$

   $$t = \frac{3.625 - 3}{2.066/\sqrt{8}} = 0.856$$

8. df = 7

   $P\text{-value} = P(t_7 > 0.856) = 0.210$

9. Since $P\text{-value} = 0.210 > 0.01$ we do not reject H_0. We do not have convincing evidence that the mean number of words recalled after 1 hour exceeds the mean number recalled after 24 hours by more than 3.

11.71

Specimen	Direct	Stratified	Difference
1	24	8	16
2	32	36	-4
3	0	8	-8
4	60	56	4
5	20	52	-32
6	64	64	0
7	40	28	12
8	8	8	0
9	12	8	4
10	92	100	-8
11	4	0	4
12	68	56	12
13	76	68	8
14	24	52	-28
15	32	28	4
16	0	0	0
17	36	36	0
18	16	12	4
19	92	92	0
20	4	12	-8
21	40	48	-8
22	24	24	0
23	0	0	0
24	8	12	-4
25	12	40	-28
26	16	12	4
27	40	76	-36

1. μ_d = mean difference in number of seeds detected (direct − stratified)
2. H_0: $\mu_d = 0$
3. H_a: $\mu_d \neq 0$
4. $\alpha = 0.05$
5. $t = \dfrac{\bar{x}_d - \text{hypothesized value}}{s_d/\sqrt{n}}$
6.

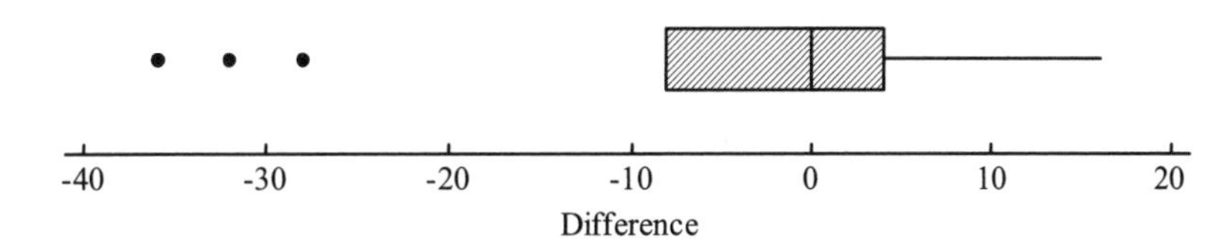

The boxplot shows a distribution of differences that is negatively skewed and has three outliers, and so the assumption that the population distribution of differences is normal is dubious. Nonetheless we will proceed with caution. Additionally, we need to assume that this set of 27 soil samples forms a random sample from the population of soil samples.

7. $\bar{x}_d = -3.407,\ s_d = 13.253$

   $t = \dfrac{-3.407 - 0}{13.253/\sqrt{27}} = -1.336$
8. df = 26

   $P\text{-value} = 2 \cdot P(t_{26} < -1.336) = 0.193$
9. Since $P\text{-value} = 0.193 > 0.05$ we do not reject H_0. We do not have convincing evidence that the mean number of seeds detected differs for the two methods.

11.73

1. p_1 = proportion of high school seniors exposed to the drug program who use marijuana

   p_2 = proportion of high school seniors not exposed to the drug program who use marijuana
2. H_0: $p_1 - p_2 = 0$
3. H_a: $p_1 - p_2 < 0$
4. $\alpha = 0.05$
5. $z = \dfrac{\hat{p}_1 - \hat{p}_2}{\sqrt{\dfrac{\hat{p}_c(1-\hat{p}_c)}{n_1} + \dfrac{\hat{p}_c(1-\hat{p}_c)}{n_2}}}$
6. We are told that the samples were random samples from the populations. Also $n_1\hat{p}_1 = 288(141/288) = 141 \geq 10,\ n_1(1-\hat{p}_1) = 288(147/288) = 147 \geq 10,$ $n_2\hat{p}_2 = 335(181/335) = 181 \geq 10,$ and $n_2(1-\hat{p}_2) = 335(154/335) = 154 \geq 10,$ so the samples are large enough.
7. $\hat{p}_c = \dfrac{141+181}{288+335} = \dfrac{322}{623}$

   $z = \dfrac{141/288 - 181/335}{\sqrt{\dfrac{(322/623)(301/623)}{288} + \dfrac{(322/623)(301/623)}{335}}} = -1.263$
8. $P\text{-value} = P(Z < -1.263) = 0.103$

9. Since P-value $= 0.103 > 0.05$ we do not reject H_0. We do not have convincing evidence that the proportion using marijuana is lower for students exposed to the DARE program.

11.75 Check of Conditions

We are told that the samples were random samples from the two communities. Also, $n_1\hat{p}_1 = 119(67/119) = 67 \geq 10$, $n_1(1-\hat{p}_1) = 119(52/119) = 52 \geq 10$, $n_2\hat{p}_2 = 143(106/143) = 106 \geq 10$, and $n_2(1-\hat{p}_2) = 143(37/143) = 37 \geq 10$, so the samples are large enough.

Calculation

The 90% confidence interval for $p_1 - p_2$ is

$$(\hat{p}_1 - \hat{p}_2) \pm (z \text{ critical value})\sqrt{\frac{\hat{p}_1(1-\hat{p}_1)}{n_1} + \frac{\hat{p}_2(1-\hat{p}_2)}{n_2}}$$

$$= \left(\frac{67}{119} - \frac{106}{143}\right) \pm 1.645\sqrt{\frac{(67/119)(52/119)}{119} + \frac{(106/143)(37/143)}{143}}$$

$$= (\mathbf{-0.274, -0.082})$$

Interpretation of Interval

We are 90% confident that $p_1 - p_2$ lies between −0.274 and −0.082, where p_1 is the proportion of children in the community with fluoridated water who have decayed teeth and p_2 is the proportion of children in the community without fluoridated water who have decayed teeth.

The interval does not contain zero, which means that we have evidence at the 0.1 level of a difference between the proportions of children with decayed teeth in the two communities, and evidence at the 0.05 level that the proportion of children with decayed teeth is smaller in the community with fluoridated water.

11.77 **a** Check of Conditions

We are told to assume that the peak loudness distributions are approximately normal, and that the participants were randomly assigned to the conditions.

Calculation

df = 17.276. The 95% confidence interval for $\mu_1 - \mu_2$ is

$$(\bar{x}_1 - \bar{x}_2) \pm (t \text{ critical value})\sqrt{\frac{s_1^2}{n_1} + \frac{s_2^2}{n_2}}$$

$$= (63 - 54) \pm 2.107\sqrt{\frac{13^2}{10} + \frac{16^2}{10}}$$

$$= (\mathbf{-4.738, 22.738})$$

Interpretation

We are 95% confident that the difference in mean loudness for open mouthed and closed mouthed eating of potato chips is between −4.738 and 22.738.

b

1. μ_1 = mean loudness for potato chips (closed-mouth chewing)
   μ_2 = mean loudness for tortilla chips (closed-mouth chewing)
2. H_0: $\mu_1 - \mu_2 = 0$
3. H_a: $\mu_1 - \mu_2 \neq 0$
4. $\alpha = 0.01$

5. $t = \dfrac{(\bar{x}_1 - \bar{x}_2) - (\text{hypothesized value})}{\sqrt{\dfrac{s_1^2}{n_1} + \dfrac{s_2^2}{n_2}}} = \dfrac{(\bar{x}_1 - \bar{x}_2) - 0}{\sqrt{\dfrac{s_1^2}{n_1} + \dfrac{s_2^2}{n_2}}}$
6. We are told to assume that the peak loudness distributions are approximately normal, and that the participants were randomly assigned to the conditions.
7. $t = \dfrac{54 - 53}{\sqrt{\dfrac{16^2}{10} + \dfrac{16^2}{10}}} = 0.140$
8. df = 18
   $P\text{-value} = 2 \cdot P(t_{18} > 0.140) = 0.890$
9. Since $P\text{-value} = 0.890 > 0.01$ we do not reject H_0. We do not have convincing evidence of a difference between potato chips and tortilla chips with respect to mean peak loudness (closed-mouth chewing).

c

1. μ_1 = mean loudness for stale tortilla chips (closed-mouth chewing)
   μ_2 = mean loudness for fresh tortilla chips (closed-mouth chewing)
2. H_0: $\mu_1 - \mu_2 = 0$
3. H_a: $\mu_1 - \mu_2 < 0$
4. $\alpha = 0.05$
5. $t = \dfrac{(\bar{x}_1 - \bar{x}_2) - (\text{hypothesized value})}{\sqrt{\dfrac{s_1^2}{n_1} + \dfrac{s_2^2}{n_2}}} = \dfrac{(\bar{x}_1 - \bar{x}_2) - 0}{\sqrt{\dfrac{s_1^2}{n_1} + \dfrac{s_2^2}{n_2}}}$
6. We are told to assume that the peak loudness distributions are approximately normal, and that the participants were randomly assigned to the conditions.
7. $t = \dfrac{53 - 56}{\sqrt{\dfrac{16^2}{10} + \dfrac{14^2}{10}}} = -0.446$
8. df = 17.688
   $P\text{-value} = P(t_{17.688} < -0.446) = 0.330$
9. Since $P\text{-value} = 0.330 > 0.05$ we do not reject H_0. We do not have convincing evidence that fresh tortilla chips are louder than stale tortilla chips.

11.79 a

1. μ_d = mean difference in systolic blood pressure between dental setting and medical setting (dental − medical)
2. H_0: $\mu_d = 0$
3. H_a: $\mu_d > 0$
4. $\alpha = 0.01$
5. $t = \dfrac{\bar{x}_d - \text{hypothesized value}}{s_d / \sqrt{n}}$
6. We need to assume that the subjects formed a random sample of patients. With this assumption, since $n = 60 \geq 30$, we can proceed with the paired t test.
7. $t = \dfrac{4.47 - 0}{8.77/\sqrt{60}} = 3.948$

8. df = 59
   $P\text{-value} = P(t_{59} > 3.948) \approx 0$
9. Since $P\text{-value} \approx 0 < 0.01$ we reject H_0. We have strong evidence that the mean blood pressure is higher in a dental setting than in a medical setting.

b 1. μ_d = mean difference in pulse rate between dental setting and medical setting (dental − medical)
2. H_0: $\mu_d = 0$
3. H_a: $\mu_d \neq 0$
4. $\alpha = 0.05$
5. $t = \dfrac{\bar{x}_d - \text{hypothesized value}}{s_d/\sqrt{n}}$
6. We need to assume that the subjects formed a random sample of patients. With this assumption, since $n = 60 \geq 30$, we can proceed with the paired t test.
7. $t = \dfrac{-1.33 - 0}{8.84/\sqrt{60}} = -1.165$
8. df = 59
   $P\text{-value} = 2 \cdot P(t_{59} < -1.165) = 0.249$
9. Since $P\text{-value} = 0.249 > 0.05$ we do not reject H_0. We do not have convincing evidence that the mean pulse rate in a dental setting is different from the mean pulse rate in a medical setting.

11.81 1. p_1 = proportion of adults who were born deaf who remove the implant
   p_2 = proportion of adults who became deaf after learning to speak who remove the implant
2. H_0: $p_1 - p_2 = 0$
3. H_a: $p_1 - p_2 \neq 0$
4. $\alpha = 0.01$
5. $z = \dfrac{\hat{p}_1 - \hat{p}_2}{\sqrt{\dfrac{\hat{p}_c(1-\hat{p}_c)}{n_1} + \dfrac{\hat{p}_c(1-\hat{p}_c)}{n_2}}}$
6. We need to assume that the samples were independent random samples from the populations. Also, $n_1\hat{p}_1 = 250(75/250) = 75 \geq 10$, $n_1(1-\hat{p}_1) = 250(175/250) = 175 \geq 10$, $n_2\hat{p}_2 = 250(25/250) = 25 \geq 10$, and $n_2(1-\hat{p}_2) = 250(225/250) = 225 \geq 10$, so the samples are large enough.
7. $\hat{p}_c = \dfrac{75+25}{250+250} = 0.2$
   $$z = \frac{75/250 - 25/250}{\sqrt{\dfrac{(0.2)(0.8)}{250} + \dfrac{(0.2)(0.8)}{250}}} = 5.590$$
8. $P\text{-value} = 2 \cdot P(Z > 5.590) \approx 0$
9. Since $P\text{-value} \approx 0 < 0.01$ we reject H_0. We have convincing evidence that the proportion of adults who were born deaf who remove the implant is different from the proportion of adults who became deaf after learning to speak who remove the implant.

Chapter 12
The Analysis of Categorical Data and Goodness-of-Fit Tests

Note: In this chapter, numerical answers were found using values from a calculator. Students using statistical tables will find that their answers differ slightly from those given.

12.1 **a** $P\text{-value} = P(\chi_2^2 > 7.5) = \mathbf{0.024}$. H_0 is not rejected.

b $P\text{-value} = P(\chi_6^2 > 13.0) = \mathbf{0.043}$. H_0 is not rejected.

c $P\text{-value} = P(\chi_9^2 > 18.0) = \mathbf{0.035}$. H_0 is not rejected.

d $P\text{-value} = P(\chi_4^2 > 21.3) = \mathbf{0.0002}$. H_0 is rejected.

e $P\text{-value} = P(\chi_3^2 > 5.0) = \mathbf{0.172}$. H_0 is not rejected.

12.3 **a** The expected counts are 80, 60 40, and 20, which are all greater than or equal to 5, so the chi-square test can be used. $P\text{-value} = P(\chi_3^2 > 19.0) = 0.0002 < 0.001$, so H_0 is rejected. We have convincing evidence that the proportions of the four types of nut are not as they are supposed to be.

b The smallest expected count would be 40(0.1) = 4, which is less than 5. So the chi-square test would not be appropriate.

12.5

Ethnicity	African-American	Asian	Caucasian	Hispanic
Observed Count	57	11	330	6
Expected Count	71.508	12.928	296.536	23.028

1. Let p_1, p_2, p_3, and p_4 be the proportions of appearances of the four ethnicities across all commercials.
2. H_0: $p_1 = 0.177$, $p_2 = 0.032$, $p_3 = 0.734$, $p_4 = 0.057$
3. H_a: H_0 is not true
4. $\alpha = 0.01$
5. $$X^2 = \sum_{\text{all cells}} \frac{(\text{observed cell count} - \text{expected cell count})^2}{\text{expected cell count}}$$
6. We need to assume that the set of commercials included in the study form a random sample from the population of commercials. All the expected counts are greater than 5.
7. $$X^2 = \frac{(57-71.508)^2}{71.508} + \cdots + \frac{(6-23.028)^2}{23.028} = 19.599$$
8. df = 3
   $P\text{-value} = P(\chi_3^2 > 19.599) \approx 0$
9. Since $P\text{-value} \approx 0 < 0.01$ we reject H_0. We have convincing evidence that the proportions of appearances in commercials are not the same as the census proportions.

12.7

Tar Level	Observed Count	Expected Count
0–7 mg	103	298.5
8–14 mg	378	298.5
15–21 mg	563	298.5
≥22 mg	150	298.5

1. Let p_1, p_2, p_3, and p_4 be the proportions of all male smoker lung cancer deaths for smokers of cigarettes of the given tar levels.
2. H_0: $p_1 = 0.25,\ p_2 = 0.25,\ p_3 = 0.25,\ p_4 = 0.25$
3. H_a: H_0 is not true
4. $\alpha = 0.05$
5. $X^2 = \sum_{\text{all cells}} \frac{(\text{observed cell count} - \text{expected cell count})^2}{\text{expected cell count}}$
6. We are told to regard the sample as representative of male smokers who die of lung cancer, so it is reasonable to treat the sample as a random sample from that population. All the expected counts are greater than 5.
7. $X^2 = \frac{(103 - 298.5)^2}{298.5} + \cdots + \frac{(150 - 298.5)^2}{298.5} = 457.464$
8. df = 3
   $P\text{-value} = P(\chi_3^2 > 457.464) \approx 0$
9. Since $P\text{-value} \approx 0 < 0.01$ we reject H_0. We have convincing evidence that the proportion of male smoker lung cancer deaths is not the same for the four given tar level categories.

12.9 **a**

Time of Day	Observed Count	Expected Count
Midnight to 3 a.m.	38	89.375
3 a.m. to 6 a.m.	29	89.375
6 a.m. to 9 a.m.	66	89.375
9 a.m. to Noon	77	89.375
Noon to 3 p.m.	99	89.375
3 p.m. to 6 p.m.	127	89.375
6 p.m. to 9 p.m.	166	89.375
9 p.m. to Midnight	113	89.375

1. Let $p_1, \ldots, p_8$ be the proportions of fatal bicycle accidents occurring in the given time periods.
2. H_0: $p_1 = p_2 = \cdots = p_8 = 0.125$
3. H_a: H_0 is not true
4. $\alpha = 0.05$
5. $X^2 = \sum_{\text{all cells}} \frac{(\text{observed cell count} - \text{expected cell count})^2}{\text{expected cell count}}$
6. We are told to regard the 715 accidents included in the study as a random sample from the population of fatal bicycle accidents. All the expected counts are greater than 5.

7. $X^2 = \frac{(38-89.375)^2}{89.375} + \cdots + \frac{(113-89.375)^2}{89.375} = 166.958$
8. df = 7

   $P\text{-value} = P(\chi_7^2 > 166.958) \approx 0$
9. Since $P\text{-value} \approx 0 < 0.05$ we reject H_0. We have convincing evidence that fatal bicycle accidents are not equally likely to occur in each of the 3-hour time periods given.

b

Time of Day	Observed Count	Expected Count
Midnight to Noon	210	238.333
Noon to Midnight	505	476.667

1. Let p_1 and p_2 be the proportions of fatal bicycle accidents occurring between midnight and noon and between noon and midnight, respectively.
2. H_0: $p_1 = 1/3,\ p_2 = 2/3$
3. H_a: H_0 is not true
4. $\alpha = 0.05$
5. $X^2 = \sum_{\text{all cells}} \frac{(\text{observed cell count} - \text{expected cell count})^2}{\text{expected cell count}}$
6. We are told to regard the 715 accidents included in the study as a random sample from the population of fatal bicycle accidents. Both of the expected counts are greater than 5.
7. $X^2 = \frac{(210-238.333)^2}{238.333} + \frac{(505-476.667)^2}{476.667} = 5.052$
8. df = 1

   $P\text{-value} = P(\chi_1^2 > 5.052) = 0.025$
9. Since $P\text{-value} = 0.025 < 0.05$ we reject H_0. Using a 0.05 significance level, we have convincing evidence that fatal bicycle accidents do not occur as stated in the hypothesis.

12.11

Age	Observed Count	Expected Count
18–34	36	70
35–64	130	102
65 and over	34	28

1. Let p_1, p_2, and p_3 be the proportions of lottery ticket purchasers who fall into the given age catergories.
2. H_0: $p_1 = 0.35,\ p_2 = 0.51,\ p_3 = 0.14$
3. H_a: H_0 is not true
4. $\alpha = 0.05$
5. $X^2 = \sum_{\text{all cells}} \frac{(\text{observed cell count} - \text{expected cell count})^2}{\text{expected cell count}}$
6. We are told to assume that the 200 people in the study form a random sample of lottery ticket purchasers. All the expected counts are greater than 5.
7. $X^2 = \frac{(36-70)^2}{70} + \cdots + \frac{(34-28)^2}{28} = 25.486$

8. df = 2

   P-value $= P(\chi_2^2 > 25.486) \approx 0$
9. Since $P\text{-value} \approx 0 < 0.05$ we reject H_0. We have convincing evidence that one or more of these three age groups buys a disproportionate share of lottery tickets.

12.13

Phenotype	1	2	3	4
Observed Count	926	288	293	104
Expected Count	906.1875	302.0625	302.0625	100.6875

1. Let p_1, p_2, p_3, and p_4 be the proportions of phenotypes resulting from the given process.
2. H_0: $p_1 = 9/16,\ p_2 = 3/16,\ p_3 = 3/16,\ p_4 = 1/16$
3. H_a: H_0 is not true
4. $\alpha = 0.01$
5. $X^2 = \sum_{\text{all cells}} \frac{(\text{observed cell count} - \text{expected cell count})^2}{\text{expected cell count}}$
6. We need to assume that the plants included in the study form a random sample from the population of such plants. All the expected counts are greater than 5.
7. $X^2 = \frac{(926-906.1875)^2}{906.1875} + \cdots + \frac{(104-100.6875)^2}{100.6875} = 1.469$
8. df = 3

   P-value $= P(\chi_3^2 > 1.469) = 0.690$
9. Since $P\text{-value} = 0.690 > 0.01$ we do not reject H_0. We do not have convincing evidence that the data from this experiment are not consistent with Mendel's laws.

12.15 **a** df $= (4-1)(5-1) = 12$. P-value $= P(\chi_{12}^2 > 7.2) = 0.844$. Since the P-value is greater than 0.1, we do not have convincing evidence that education level and preferred candidate are not independent.

b df $= (4-1)(4-1) = 9$. P-value $= P(\chi_9^2 > 14.5) = 0.106$. Since the P-value is greater than 0.05, we do not have convincing evidence that education level and preferred candidate are not independent.

12.17

	Body Piercings Only	**Tattoos Only**	**Both Body Piercing and Tattoos**	**No Body Art**
Freshman	61 (49.714)	7 (15.086)	14 (18.514)	86 (84.686)
Sophomore	43 (37.878)	11 (11.494)	10 (14.106)	64 (64.522)
Junior	20 (23.378)	9 (7.094)	7 (8.706)	43 (39.822)
Senior	21 (34.031)	17 (10.327)	23 (12.673)	54 (57.969)

H_0: Class standing and body art response are independent
H_a: Class standing and body art response are not independent
$\alpha = 0.01$

$$X^2 = \sum_{\text{all cells}} \frac{(\text{observed cell count} - \text{expected cell count})^2}{\text{expected cell count}}$$

We are told to regard the sample as representative of the students at this university, so we are justified in treating it as a random sample from that population. All the expected counts are greater than 5.

$$X^2 = \frac{(61-49.714)^2}{49.714} + \cdots + \frac{(54-57.969)^2}{57.969} = 29.507$$

df = 9

P-value $= P(\chi_9^2 > 29.507) = 0.001$

Since $P\text{-value} = 0.001 < 0.01$ we reject H_0. We have convincing evidence of an association between class standing and response to the body art question.

12.19 **a** H_0: Field of study and smoking status are independent
H_a: Field of study and smoking status are not independent
$\alpha = 0.01$

$$X^2 = \sum_{\text{all cells}} \frac{(\text{observed cell count} - \text{expected cell count})^2}{\text{expected cell count}}$$

We are told that the sample was a random sample from the population. All the expected counts are greater than 5.

$X^2 = 90.853$
df = 8
P-value ≈ 0
Since $P\text{-value} \approx 0 < 0.01$ we reject H_0. We have convincing evidence that smoking status and field of study are not independent.

b The particularly high contributions to the chi-square statistic (in order of importance) come from the field of communication, languages, and cultural studies, where there was a disproportionately high number of smokers, the field of mathematics, engineering, and sciences, where there was a disproportionately low number of smokers, and the field of social science and human services, where there was a disproportionately high number of smokers.

12.21 **a**

	Usually Eat 3 Meals a Day	**Rarely Eat 3 Meals a Day**
Male	26 (21.755)	22 (26.245)
Female	37 (41.245)	54 (49.755)

H_0: The proportions falling into the two response categories are the same for males and females.
H_a: The proportions falling into the two response categories are not the same for males and females.
$\alpha = 0.05$

$$X^2 = \sum_{\text{all cells}} \frac{(\text{observed cell count} - \text{expected cell count})^2}{\text{expected cell count}}$$

We are told to assume that the samples of male and female students were random samples from the populations. All the expected counts are greater than 5.

$$X^2 = \frac{(26-21.755)^2}{21.755} + \cdots + \frac{(54-49.755)^2}{49.755} = 2.314$$

df = 1

$P\text{-value} = P(\chi_1^2 > 2.314) = 0.128$

Since $P\text{-value} = 0.128 > 0.05$ we do not reject H_0. We do not have convincing evidence that the proportions falling into the two response categories are not the same for males and females.

b Yes.

c Yes. Since $P\text{-value} = 0.127 > 0.05$ we do not reject H_0. We do not have convincing evidence that the proportions falling into the two response categories are not the same for males and females.

d The two P-values are almost equal, in fact the difference between them is only due to rounding errors in the MINITAB program. In other words, if complete accuracy had been maintained throughout, the two P-values would have been exactly equal. (Also, the chi-square statistic in Part (a) is the square of the z statistic in Part (c).) It should not be surprising that the P-values are at least similar, since both measure the probability of getting sample proportions at least as far from the expected proportions, given that the proportions who usually eat three meals per day are the same for the two populations.

12.23 a

	Donation	**No Donation**
No Gift	397 (514.512)	2865 (2747.488)
Small Gift	465 (510.569)	2772 (2726.431)
Large Gift	691 (527.919)	2656 (2819.081)

H_0: The proportions falling into the two donation categories are the same for all three gift treatments.
H_a: The proportions falling into the two donation categories are not the same for all three gift treatments.
$\alpha = 0.01$

$$X^2 = \sum_{\text{all cells}} \frac{(\text{observed cell count} - \text{expected cell count})^2}{\text{expected cell count}}$$

We are told that the three treatments were assigned at random. All the expected counts are greater than 5.

$$X^2 = \frac{(397 - 514.512)^2}{514.512} + \cdots + \frac{(2656 - 2819.081)^2}{2819.081} = 96.506$$

df = 2

$P\text{-value} = P(\chi_2^2 > 96.506) \approx 0$

Since $P\text{-value} \approx 0 < 0.01$ we reject H_0. We have convincing evidence that the proportions falling into the two donation categories are not the same for all three gift treatments.

b The result of Part (a) tells us that the level of the gift seems to make a difference. Looking at the data given, 12% of those receiving no gift made a donation, 14% of those receiving a small gift made a donation, and 21% of those receiving a large gift made a donation. (These percentages can be compared to 16% making donations amongst the expected counts.) So it seems that the most effective strategy is to include a large gift, with the small gift making very little difference compared to no gift at all.

12.25

		Alcohol Exposure Group			
		1	**2**	**3**	**4**
School Performance	**Excellent**	110 (79.25)	93 (79.25)	49 (79.25)	65 (79.25)
	Good	328 (316)	325 (316)	316 (316)	295 (316)
	Average/Poor	239 (281.75)	259 (281.75)	312 (281.75)	317 (281.75)

H_0: Alcohol exposure and school performance are independent
H_a: Alcohol exposure and school performance are not independent
$\alpha = 0.05$

$$X^2 = \sum_{\text{all cells}} \frac{(\text{observed cell count} - \text{expected cell count})^2}{\text{expected cell count}}$$

We are told to regard the sample as a random sample of German adolescents. All the expected counts are greater than 5.

$$X^2 = \frac{(110 - 79.25)^2}{79.25} + \cdots + \frac{(317 - 281.75)^2}{281.75} = 46.515$$

df = 6

$P\text{-value} = P(\chi_6^2 > 46.515) \approx 0$

Since $P\text{-value} \approx 0 < 0.05$ we reject H_0. We have convincing evidence of an association between alcohol exposure and school performance.

12.27

Number of Sweet Drinks Consumed per Day	Overweight? Yes	No
0	22 (28.921)	930 (923.079)
1	73 (65.225)	2074 (2081.775)
2	56 (52.769)	1681 (1684.231)
3 or More	102 (106.085)	3390 (3385.915)

H_0: Number of sweet drinks consumed per day and weight status are independent
H_a: Number of sweet drinks consumed per day and weight status are not independent
$\alpha = 0.05$

$$X^2 = \sum_{\text{all cells}} \frac{(\text{observed cell count} - \text{expected cell count})^2}{\text{expected cell count}}$$

We are told to regard the sample as representative of 2- to 3-year-old children, so we are justified in treating it as a random sample from that population. All the expected counts are greater than 5.

$$X^2 = \frac{(22 - 28.921)^2}{28.921} + \cdots + \frac{(3390 - 3385.915)^2}{3385.915} = 3.030$$

df = 3

$P\text{-value} = P(\chi_3^2 > 3.030) = 0.387$

Since $P\text{-value} = 0.387 > 0.05$ we do not reject H_0. We do not have convincing evidence of an association between whether or not children are overweight after one year and the number of sweet drinks consumed.

12.29

	City		
Vehicle Type	**Concord**	**Pleasant Hills**	**North San Francisco**
Small	68 (89.060)	83 (107.019)	221 (175.921)
Compact	63 (56.740)	68 (68.181)	106 (112.079)
Midsize	88 (84.511)	123 (101.553)	142 (166.936)
Large	24 (12.689)	18 (15.247)	11 (25.064)

H_0: City of residence and vehicle type are independent
H_a: City of residence and vehicle type are not independent
$\alpha = 0.05$

$$X^2 = \sum_{\text{all cells}} \frac{(\text{observed cell count} - \text{expected cell count})^2}{\text{expected cell count}}$$

We are told to regard the sample as a random sample of Bay area residents. All the expected counts are greater than 5.

$$X^2 = \frac{(68-89.060)^2}{89.060} + \cdots + \frac{(11-25.064)^2}{25.064} = 49.813$$

df = 6

$P\text{-value} = P(\chi_6^2 > 49.813) \approx 0$

Since $P\text{-value} \approx 0 < 0.05$ we reject H_0. We have convincing evidence of an association between city of residence and vehicle type.

12.31

	View		
Sex ID	**Front**	**Profile**	**Three-Quarter**
Correct	23 (26)	26 (26)	29 (26)
Incorrect	17 (14)	14 (14)	11 (14)

H_0: The proportions of correct sex identifications are the same for all three nose views.
H_a: The proportions of correct sex identifications are not the same for all three nose views.
$\alpha = 0.05$

$$X^2 = \sum_{\text{all cells}} \frac{(\text{observed cell count} - \text{expected cell count})^2}{\text{expected cell count}}$$

We need to assume that the students were randomly assigned to the nose views. All the expected counts are greater than 5.

$$X^2 = \frac{(23-26)^2}{26} + \cdots + \frac{(11-14)^2}{14} = 1.978$$

df = 2

$P\text{-value} = P(\chi_2^2 > 1.978) = 0.372$

Since $P\text{-value} = 0.372 > 0.05$ we do not reject H_0. We do not have convincing evidence that the proportions of correct sex identifications are not the same for all three nose views.

12.33 **a** The number of men in the sample who napped is $744(0.38) = 282.72$, which we round to 283, since the number of men who napped must be a whole number. The number of men who did not nap is therefore $744 - 283 = 461$. The observed frequencies for the women are calculated in a similar way. (The table below also shows the expected frequencies in parentheses.)

	Napped	Did Not Nap	Row Total
Men	283 (257)	461 (487)	**744**
Women	231 (257)	513 (487)	**744**

b H_0: Gender and napping are independent
H_a: Gender and napping are not independent
$\alpha = 0.01$

$$X^2 = \sum_{\text{all cells}} \frac{(\text{observed cell count} - \text{expected cell count})^2}{\text{expected cell count}}$$

We are told that the sample was nationally representative, so we are justified in treating it as a random sample from the population of American adults. All the expected counts are greater than 5.

$$X^2 = \frac{(283-257)^2}{257} + \cdots + \frac{(513-487)^2}{487} = 8.034$$

df = 1

$P\text{-value} = P(\chi_1^2 > 8.034) = 0.005$

Since $P\text{-value} = 0.005 < 0.01$ we reject H_0. We have convincing evidence of an association between gender and napping.

c Yes. We have convincing evidence at the 0.01 significance level of an association between gender and napping in the population. This is equivalent to saying that we have convincing evidence at the 0.01 significance level that the proportions of men and women who nap are different (a two-tailed test of a difference of the proportions). Thus, converting this to a one-tailed test, since in the sample the proportion of men who napped was greater than the proportion of women who napped, we have convincing evidence at the 0.005 level that a greater proportion of men nap than women.

12.35

Day	Observed Count	Expected Count
Sunday	14	14.286
Monday	13	14.286
Tuesday	12	14.286
Wednesday	15	14.286
Thursday	14	14.286
Friday	17	14.286
Saturday	15	14.286

1. Let $p_1, \ldots, p_7$ be the proportions of all fatal bicycle accidents occurring on the seven days.
2. H_0: $p_1 = p_2 = \cdots = p_7 = 1/7$
3. H_a: H_0 is not true
4. $\alpha = 0.05$
5. $$X^2 = \sum_{\text{all cells}} \frac{(\text{observed cell count} - \text{expected cell count})^2}{\text{expected cell count}}$$

6. We are told that the 100 accidents formed a random sample from the population of fatal bicycle accidents. All the expected counts are greater than 5.
7. $X^2 = \frac{(14-14.286)^2}{14.286} + \cdots + \frac{(15-14.286)^2}{14.286} = 1.08$
8. df = 6

   $P\text{-value} = P(\chi_6^2 > 1.08) = 0.982$
9. Since $P\text{-value} = 0.982 > 0.05$ we do not reject H_0. We do not have convincing evidence that the proportion of accidents is not the same for all days of the week.

12.37

	Response				
Country	**Never**	**Rarely**	**Sometimes**	**Often**	**Not Sure**
Italy	600 (400)	140 (222)	140 (244)	90 (90)	30 (44)
Spain	540 (400)	160 (222)	140 (244)	70 (90)	90 (44)
France	400 (400)	250 (222)	200 (244)	120 (90)	30 (44)
United States	360 (400)	230 (222)	270 (244)	110 (90)	30 (44)
South Korea	100 (400)	330 (222)	470 (244)	60 (90)	40 (44)

H_0: The proportions falling into the response categories are all the same for all five countries.
H_a: The proportions falling into each of the response categories are not all the same for all five countries.
$\alpha = 0.01$

$$X^2 = \sum_{\text{all cells}} \frac{(\text{observed cell count} - \text{expected cell count})^2}{\text{expected cell count}}$$

We are told that the samples were random samples from the populations. All the expected counts are greater than 5.

$$X^2 = \frac{(600-400)^2}{400} + \cdots + \frac{(40-44)^2}{44} = 881.360$$

df = 16

$P\text{-value} = P(\chi_{16}^2 > 881.360) \approx 0$

Since $P\text{-value} \approx 0 < 0.01$ we reject H_0. We have convincing evidence that the response proportions are not all the same for all five countries.

12.39

Season	**Winter**	**Spring**	**Summer**	**Fall**
Observed Count	328	334	372	327
Expected Count	340.25	340.25	340.25	340.25

1. Let p_1, p_2, p_3, and p_4 be the proportions of homicides occurring in the four seasons.
2. H_0: $p_1 = \cdots = p_4 = 0.25$
3. H_a: H_0 is not true
4. $\alpha = 0.05$
5. $X^2 = \sum_{\text{all cells}} \frac{(\text{observed cell count} - \text{expected cell count})^2}{\text{expected cell count}}$
6. We need to assume that the 1361 homicides form a random sample from the population of homicides. All the expected counts are greater than 5.

7. $X^2 = \frac{(328-340.25)^2}{340.25} + \cdots + \frac{(327-340.25)^2}{340.25} = 4.035$
8. df = 3

   $P\text{-value} = P(\chi_3^2 > 4.035) = 0.258$
9. Since $P\text{-value} = 0.258 > 0.05$ we do not reject H_0. We do not have convincing evidence that the homicide rate is not the same over the four seasons.

12.41

	Role	
Position	**Initiate Chase**	**Participate in Chase**
Center	28 (39.038)	48 (36.962)
Wing	66 (54.962)	41 (52.038)

H_0: Position and role are independent
H_a: Position and role are not independent
$\alpha = 0.01$

$$X^2 = \sum_{\text{all cells}} \frac{(\text{observed cell count} - \text{expected cell count})^2}{\text{expected cell count}}$$

Each of the 183 observations in the sample is a particular lioness on a particular hunt. (Presumably several observations could have been gathered for a single lioness, each for a different hunt. Likewise, several observations could have been gathered from a single hunt, each for a different lioness.) We need to assume that these 183 observations form a random sample of lioness-hunts. All the expected counts are greater than 5.

$$X^2 = \frac{(28-39.038)^2}{39.038} + \cdots + \frac{(41-52.038)^2}{52.038} = 10.976$$

df = 1

$P\text{-value} = P(\chi_1^2 > 10.976) = 0.001$

Since $P\text{-value} = 0.001 < 0.01$ we reject H_0. We have convincing evidence of an association between position and role.

The required assumption is given above.

12.43

	Response	
Region	**Agree**	**Disagree**
Northeast	130 (150.350)	59 (38.650)
West	146 (149.554)	42 (38.446)
Midwest	211 (209.217)	52 (53.783)
South	291 (268.879)	47 (69.121)

H_0: Response (agree/disagree) and region of residence are independent
H_a: Response (agree/disagree) and region of residence are not independent
$\alpha = 0.01$

$$X^2 = \sum_{\text{all cells}} \frac{(\text{observed cell count} - \text{expected cell count})^2}{\text{expected cell count}}$$

We are told that the sample was a random sample of adults. All the expected counts are greater than 5.

$$X^2 = \frac{(130-150.350)^2}{150.350} + \cdots + \frac{(47-69.121)^2}{69.121} = 22.855$$

df = 3

P-value $= P(\chi_3^2 > 22.855) \approx 0$

Since P-value $\approx 0 < 0.01$ we reject H_0. We have convincing evidence of an association between response and region of residence.

12.45 a

Astrological Sign	Observed Count	Expected Count
Aquarius	35666	38347.333
Aries	37926	38347.333
Cancer	38126	38347.333
Capricorn	54906	38347.333
Gemini	37179	38347.333
Leo	37354	38347.333
Libra	37910	38347.333
Pisces	36677	38347.333
Sagittarius	34175	38347.333
Scorpio	35352	38347.333
Taurus	37179	38347.333
Virgo	37718	38347.333

1. Let $p_1, \ldots, p_{12}$ be the proportions of male insured drivers born under the twelve astrological signs.
2. H_0: $p_1 = p_2 = \cdots = p_{12} = 1/12$
3. H_a: H_0 is not true
4. $\alpha = 0.05$
5. $X^2 = \sum_{\text{all cells}} \frac{(\text{observed cell count} - \text{expected cell count})^2}{\text{expected cell count}}$
6. We are told to treat the male policyholders of this company as a random sample of male insured drivers from Australia. All the expected counts are greater than 5.
7. $X^2 = \frac{(35666-38347.333)^2}{38347.333} + \cdots + \frac{(37718-38347.333)^2}{38347.333} = 8216.476$
8. df = 11

   P-value $= P(\chi_{11}^2 > 8216.476) \approx 0$
9. Since P-value $\approx 0 < 0.05$ we reject H_0. We have convincing evidence that the proportions of male insured drivers for the twelve astrological signs are not all equal.

b This could occur if the birthrate is higher for the time of year designated as "Capricorn" than it is for other times of the year.

c The total number of policyholders listed in the first table is 460168. Therefore, for example, the proportion of policyholders born under Aquarius is $35666/460168$. The total number of claims listed in the second table is 1000. So, if the numbers of claims were in proportion to the numbers of policyholders, then we would expect the number of claims for policyholders born under Aquarius to be $1000(35666/460168) = 77.506$. This is the expected count for Aquarius, and the other expected counts are calculated in a similar way.

Astrological Sign	Observed Count	Expected Count
Aquarius	85	77.506
Aries	83	82.418
Cancer	82	82.852
Capricorn	88	119.317
Gemini	83	80.794
Leo	83	81.175
Libra	83	82.383
Pisces	82	79.703
Sagittarius	81	74.266
Scorpio	85	76.824
Taurus	84	80.794
Virgo	81	81.966

1. Let $p_1, \ldots, p_{12}$ be the proportions of all claims for drivers born under the 12 astrological signs.
2. H_0: $p_1 = 35666/460168$, $p_2 = 37926/460168$, and so on, using the data given in the first table.
3. H_a: H_0 is not true
4. $\alpha = 0.05$
5. $$X^2 = \sum_{\text{all cells}} \frac{(\text{observed cell count} - \text{expected cell count})^2}{\text{expected cell count}}$$
6. We are told that this is a random sample of claims for this company. All the expected counts are greater than 5.
7. $$X^2 = \frac{(85 - 77.506)^2}{77.506} + \cdots + \frac{(81 - 81.966)^2}{81.966} = 10.748$$
8. df = 11

   $P\text{-value} = P(\chi^2_{11} > 10.748) = 0.465$
9. Since $P\text{-value} = 0.465 > 0.05$ we do not reject H_0. We do not have convincing evidence that the proportions of claims submitted by drivers born under the twelve astrological signs are not equal to the corresponding proportions of policyholders.

Chapter 13
Simple Linear Regression and Correlation: Inferential Methods

Note: In this chapter, numerical answers to questions involving the normal, t, and chi square distributions were found using values from a calculator. Students using statistical tables will find that their answers differ slightly from those given.

13.1 **a** $y = -5.0 + 0.017x$

b When $x = 1000$, $y = -5 + 0.017(1000) = 12$.
When $x = 2000$, $y = -5 + 0.017(2000) = 29$.

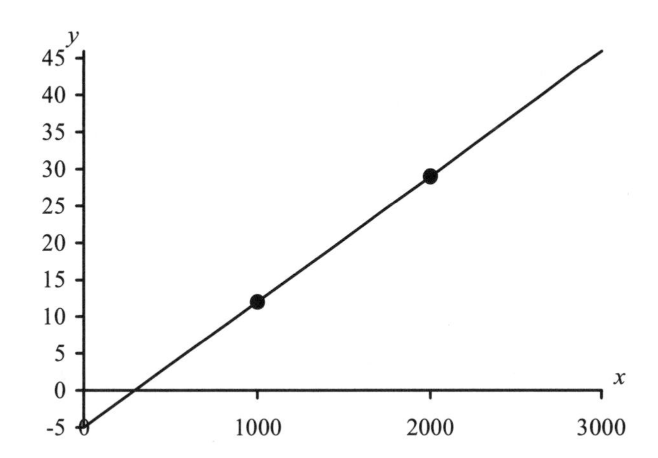

c When $x = 2100$, $y = -5 + 0.017(2100) = 30.7$. The mean gas usage for houses with 2100 square feet of space is **30.7** therms.

d **0.017** therms

e 100(0.017) = **1.7** therms

f No. The given relationship only applies to houses whose sizes are between 1000 and 3000 square feet. The size of this house, 500 square feet, lies outside this range.

13.3 **a** When $x = 15$, $\mu_y = 0.135 + 0.003(15) = \mathbf{0.18}$ micrometers
When $x = 17$, $\mu_y = 0.135 + 0.003(17) = \mathbf{0.186}$ micrometers

b When $x = 15$, μ_y=0.18, so $P(y > 0.18) = \mathbf{0.5}$.

c When $x = 14$, $\mu_y = 0.135 + 0.003(14) = 0.177$,

so $P(y > 0.175) = P\left(z > \dfrac{0.175 - 0.177}{0.005}\right) = P(z > -0.4) = \mathbf{0.655}$.

$P(y < 0.178) = P\left(z < \dfrac{0.178 - 0.177}{0.005}\right) = P(z < 0.2) = \mathbf{0.579}$.

13.5 **a** Average change in price associated with one extra square foot of space = $**47**.
Average change in price associated with 100 extra square feet of space = 100(47) = **$4700**.

b When $x = 1800,\ \mu_y = 23000 + 47(1800) = 107600$.

So $P(y > 110000) = P\left(z > \dfrac{110000 - 107600}{5000}\right) = P(z > 0.48) = \mathbf{0.316}$.

$P(y < 100000) = P\left(z < \dfrac{100000 - 107600}{5000}\right) = P(z < -1.52) = \mathbf{0.064}$.

13.7 **a** $r^2 = 1 - \dfrac{\text{SSResid}}{\text{SSTo}} = 1 - \dfrac{0.313}{0.356} = \mathbf{0.121}$.

b A point estimate of σ is $s_e = \sqrt{\dfrac{\text{SSResid}}{n-2}} = \sqrt{\dfrac{0.313}{13}} = \mathbf{0.155}$. This is a typical deviation of a bone mineral density value in the sample from the value predicted by the least-squares line.

c **0.009** g/cm^2.

d When $x = 60$, estimate of mean BMD $= 0.558 + 0.009(60) = 1.098$ g/cm^2.

13.9 **a** The required proportion is $r^2 = 1 - \dfrac{\text{SSResid}}{\text{SSTo}} = 1 - \dfrac{2620.57}{22398.05} = \mathbf{0.883}$.

b $s_e = \sqrt{\dfrac{\text{SSResid}}{n-2}} = \sqrt{\dfrac{2620.57}{14}} = \mathbf{13.682}$. The number of degrees of freedom associated with this estimate is $n - 2 = 16 - 2 = \mathbf{14}$.

13.11 **a**

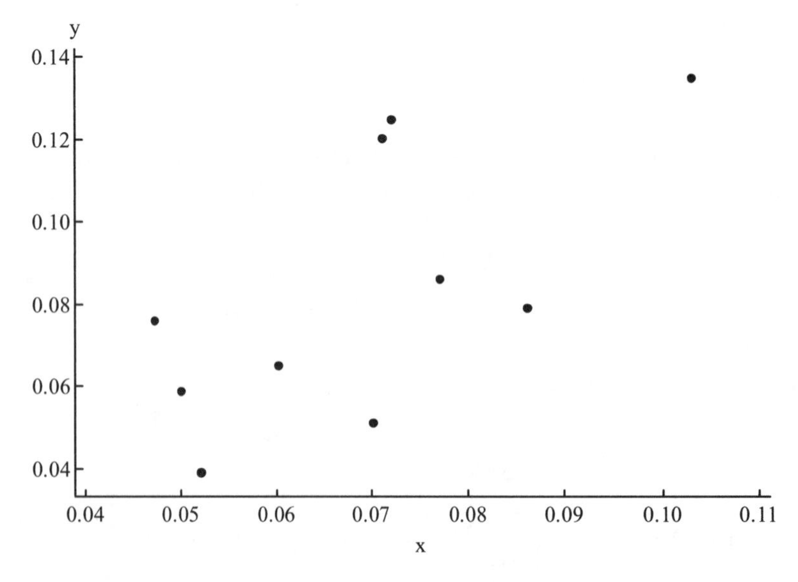

The plot shows a linear pattern, and the vertical spread of points does not appear to be changing over the range of x values in the sample. If we assume that the distribution of errors at any given x value is approximately normal, then the simple linear regression model seems appropriate.

b $\hat{y} = -0.00227 + 1.247x$
When $x = 0.09$, $\hat{y} = -0.00227 + 1.247(0.09) = \mathbf{0.110}$.

c $r^2 = \mathbf{0.436}$. This tells us that 43.6% of the variation in market share can be explained by the linear regression model relating market share and advertising share.

d A point estimate of σ is $s_e = \sqrt{\frac{\text{SSResid}}{n-2}} = \sqrt{\frac{0.00551}{8}} = \mathbf{0.026}$. The number of degrees of freedom associated with this estimate is $n - 2 = 10 - 2 = \mathbf{8}$.

13.13 **a** $S_{xx} = \sum (x - \bar{x})^2 = (5-15)^2 + \cdots + (25-15)^2 = 250$. So $\sigma_b = \frac{\sigma}{\sqrt{S_{xx}}} = \frac{4}{\sqrt{250}} = \mathbf{0.253}$.

b Now $S_{xx} = 2(250) = 500$. So $\sigma_b = \frac{\sigma}{\sqrt{S_{xx}}} = \frac{4}{\sqrt{500}} = 0.179$. No, σ_b is not half of what it was in Part (a).

c **Four** observations should be taken at each of the x values, since then S_{xx} would be multiplied by 4, and so σ_b would be divided by 2. To verify, $S_{xx} = 4(250) = 1000$, so
$\sigma_b = \frac{\sigma}{\sqrt{S_{xx}}} = \frac{4}{\sqrt{1000}} = 0.126 = (1/2)(0.253)$.

13.15 **a** $s_e = \sqrt{\frac{\text{SSResid}}{n-2}} = \sqrt{\frac{1235.470}{13}} = 9.749$.

$s_b = \frac{s_e}{\sqrt{S_{xx}}} = \frac{9.749}{\sqrt{4024.2}} = \mathbf{0.154}$.

b We must assume that the conditions for inference are met.
df = 13.
The 95% confidence interval for β is

$$b \pm (t \text{ critical value}) \cdot s_b = 2.5 \pm (2.160)(0.154) = \mathbf{(2.168, 2.832)}.$$

We are 95% confident that the slope of the population regression line relating hardness of molded plastic and time elapsed since the molding was completed is between 2.168 and 2.832.

c Yes. Since the confidence interval is relatively narrow it seems that β has been somewhat precisely estimated.

13.17 **a** $S_{xy} = \sum xy - \frac{(\sum x)(\sum y)}{n} = 44194 - \frac{(50)(16705)}{20} = 2431.5$

$S_{xx} = \sum x^2 - \frac{(\sum x)^2}{n} = 150 - \frac{(50)^2}{20} = 25.$

$\bar{x} = \frac{50}{20} = 2.5,\ \bar{y} = \frac{16705}{20} = 835.25$

The slope of the population regression line is estimated by $b = \frac{S_{xy}}{S_{xx}} = \frac{2431.5}{25} = \mathbf{97.26}.$

The y intercept of the population regression line is estimated by $a = \bar{y} - b\bar{x}$ $= 835.25 - 97.26(2.5) = \mathbf{592.1}.$

b When $x = 2,\ \hat{y} = a + bx = 592.1 + 97.26(2) = \mathbf{786.62}.$
Residual $= y - \hat{y} = 757 - 786.62 = \mathbf{-29.62}.$

c We require a 99% confidence interval for β .
We must assume that the conditions for inference are met.
$\text{SSResid} = \sum y^2 - a\sum y - b\sum xy = 14194231 - 592.1(16705) - 97.26(44194) = 4892.06.$

$s_e = \sqrt{\frac{\text{SSResid}}{n-2}} = \sqrt{\frac{4892.06}{18}} = 16.486.$

$s_b = \frac{s_e}{\sqrt{S_{xx}}} = \frac{16.486}{\sqrt{25}} = 3.297.$

df = 18.
The 99% confidence interval for β is

$$b \pm (t \text{ critical value}) \cdot s_b = 97.26 \pm (2.878)(3.297) = (\mathbf{87.769, 106.751})$$

We are 99% confident that the slope of the population regression line relating amount of oxygen consumed and time spent exercising is between 87.769 and 106.751.

13.19
1. β = slope of the population regression line relating brain volume change with mean childhood blood lead level.
2. H_0: $\beta = 0$
3. H_a: $\beta \neq 0$
4. $\alpha = 0.05$
5. $t = \frac{b - (\text{hypothesized value})}{s_b} = \frac{b - 0}{s_b}$
6. We are told to assume that the basic assumptions of the simple linear regression model are reasonably met.
7. $t = -3.66$
8. P-value ≈ 0
9. Since $P\text{-value} \approx 0 < 0.05$ we reject H_0. We have convincing evidence that the slope of the population regression line relating brain volume change with mean childhood blood lead level is not equal to zero, that is, that there is a useful linear relationship between these two variables.

13.21 **a** For the data given, $b = 0.140$, $s_e = 0.402$, $s_b = 0.026$.
The data are plotted in the scatterplot below.

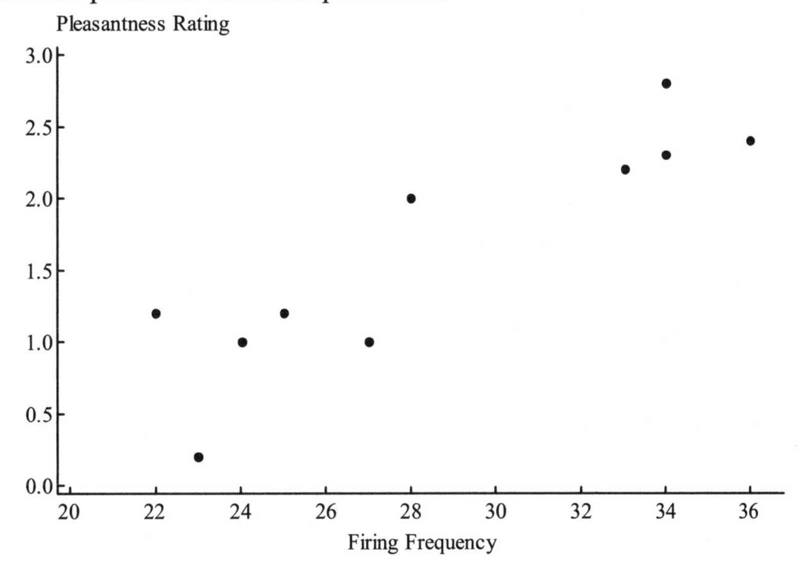

The plot shows a linear pattern, and the vertical spread of points does not appear to be changing over the range of x values in the sample. If we assume that the distribution of errors at any given x value is approximately normal, then the simple linear regression model seems appropriate.

df = 8. The 95% confidence interval for β is

$$b \pm (t \text{ critical value}) \cdot s_b = 0.140 \pm (2.306)(0.026) = (\mathbf{0.081, 0.199}).$$

We are 95% confident that the mean change in pleasantness rating associated with an increase of 1 impulse per second in firing frequency is between 0.081 and 0.199.

b

1. β = slope of the population regression line relating pleasantness rating to firing frequency.
2. H_0: $\beta = 0$
3. H_a: $\beta \neq 0$
4. $\alpha = 0.05$
5. $t = \dfrac{b - (\text{hypothesized value})}{s_b} = \dfrac{b - 0}{s_b}$
6. The conditions for inference were checked in Part (a).
7. $t = \dfrac{0.140 - 0}{0.026} = 5.451$
8. df = 8
   $P\text{-value} = 2 \cdot P(t_8 > 5.451) = 0.001$
9. Since $P\text{-value} = 0.001 < 0.05$ we reject H_0. We have convincing evidence of a useful linear relationship between firing frequency and pleasantness rating.

13.23 **a**

1. β = average change in sales revenue associated with a 1-unit increase in advertising expenditure.
2. H_0: $\beta = 0$

3. H_a: $\beta \neq 0$
4. $\alpha = 0.05$
5. $t = \dfrac{b - (\text{hypothesized value})}{s_b} = \dfrac{b-0}{s_b}$
6. We must assume that the conditions for inference are met.
7. $t = \dfrac{52.27-0}{8.05} = 6.493$
8. df = 13
   $P\text{-value} = 2 \cdot P(t_{13} > 6.493) \approx 0$
9. Since $P\text{-value} \approx 0 < 0.05$ we reject H_0. We have convincing evidence that the slope of the population regression line relating sales revenue and advertising expenditure is not equal to zero.

We conclude that there is a useful linear relationship between sales revenue and advertising expenditure.

b
1. β = average change in sales revenue associated with a 1-unit increase in advertising expenditure.
2. H_0: $\beta = 40$
3. H_a: $\beta > 40$
4. $\alpha = 0.01$
5. $t = \dfrac{b - (\text{hypothesized value})}{s_b} = \dfrac{b-40}{s_b}$
6. We must assume that the conditions for inference are met.
7. $t = \dfrac{52.27-40}{8.05} = 1.524$
8. df = 13
   $P\text{-value} = P(t_{13} > 1.524) = 0.076$
9. Since $P\text{-value} = 0.076 > 0.01$ we do not reject H_0. We do not have convincing evidence that the average change in sales revenue associated with a 1-unit (that is, \$1000) increase in advertising expenditure is greater than \$40,000.

13.25 $S_{xy} = \sum xy - \dfrac{(\sum x)(\sum y)}{n} = 4376.36 - \dfrac{(678)(104.54)}{16} = -53.5225.$

$$S_{xx} = \sum x^2 - \frac{(\sum x)^2}{n} = 36056 - \frac{(678)^2}{16} = 7325.75.$$

$$\bar{x} = \frac{678}{16} = 42.375,\ \bar{y} = \frac{104.54}{16} = 6.53375$$

$$b = \frac{S_{xy}}{S_{xx}} = \frac{-53.5225}{7325.75} = -0.00731.$$

$$a = \bar{y} - b\bar{x} = 6.53375 - (-0.00731)(42.375) = 6.843.$$

$$\text{SSResid} = \sum y^2 - a\sum y - b\sum xy = 36056 - (6.843)(104.54) - (-0.00731)(4376.36) = 0.01774$$

$$s_e = \sqrt{\frac{\text{SSResid}}{n-2}} = \sqrt{\frac{0.01774}{14}} = 0.03559.$$

$$s_b = \frac{s_e}{\sqrt{S_{xx}}} = \frac{0.03559}{\sqrt{7325.75}} = 0.000416.$$

1. β = average change in milk pH associated with a 1-unit increase in temperature.
2. H_0: $\beta = 0$
3. H_a: $\beta < 0$
4. $\alpha = 0.01$
5. $t = \dfrac{b - (\text{hypothesized value})}{s_b} = \dfrac{b-0}{s_b}$
6. The data are plotted in the scatterplot below.

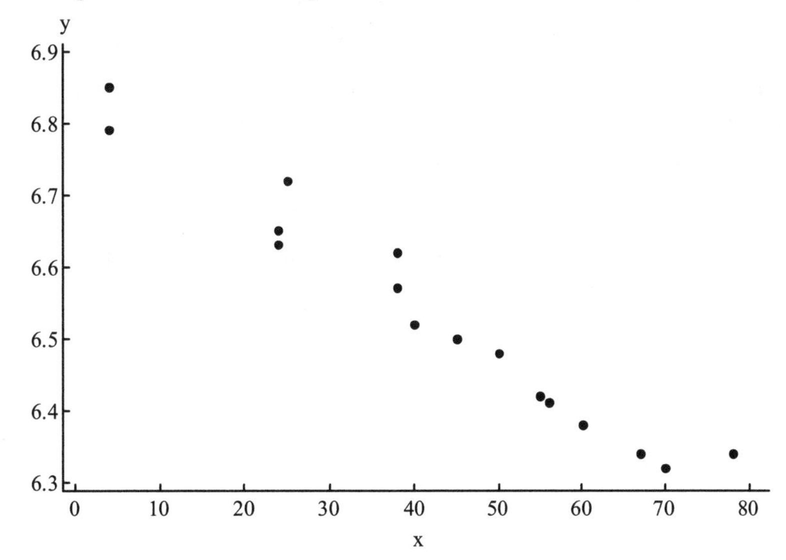

The plot shows a linear pattern, and the vertical spread of points does not appear to be changing over the range of x values in the sample. If we assume that the distribution of errors at any given x value is approximately normal, then the simple linear regression model seems appropriate.

7. $t = \dfrac{-0.00731 - 0}{0.000416} = -17.569$
8. df = 14

   P-value $= P(t_{14} < -17.569) \approx 0$
9. Since P-value $\approx 0 < 0.01$ we reject H_0. We have convincing evidence that the slope of the population regression line relating milk pH and temperature is negative. Thus the data strongly suggest that there is a negative linear relationship between temperature and pH.

13.27 a

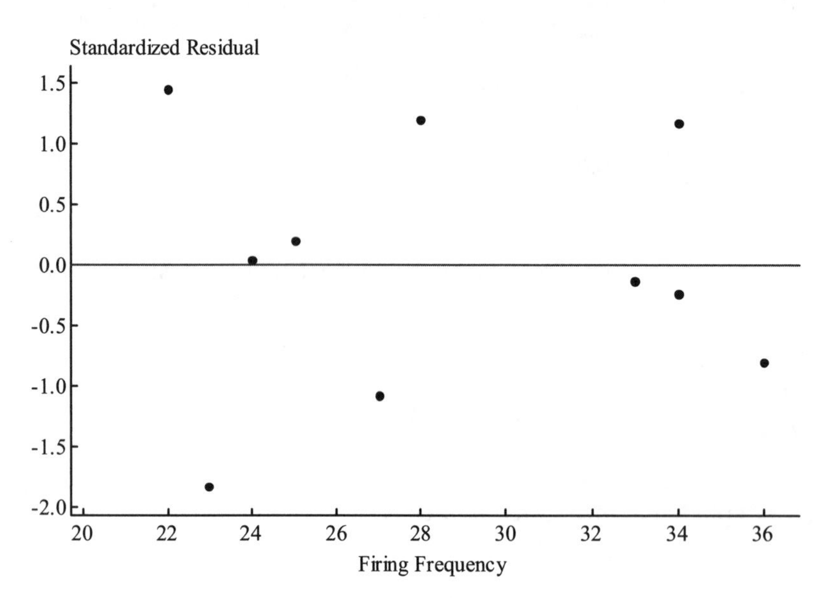

There are no particularly unusual features in the standardized residual plot. The only slightly unusual feature is the point whose standardized residual is -1.83, which is relatively far from zero, but not particularly extreme. The plot supports the assumption that the simple linear regression model applies.

b Yes. Since the normal probability plot shows a roughly linear pattern we can conclude that is it reasonable to assume that the error distribution is approximately normal.

13.29 a Letting x = minimum width and y = maximum width, the least-squares regression line is $\hat{y} = 0.939 + 0.873x$.

b The residuals and the standardized residuals are shown in the table below.

Product	Minimum Width	Maximum Width	Residual	Standardized Residual
1	1.8	2.5	-0.011	-0.016
2	2.7	2.9	-0.397	-0.601
3	2	2.15	-0.535	-0.816
4	2.6	2.9	-0.309	-0.469
5	3.15	3.2	-0.489	-0.742
6	1.8	2	-0.511	-0.780
7	1.5	1.6	-0.649	-0.996
8	3.8	4.8	0.543	0.825
9	5	5.9	0.595	0.918
10	4.75	5.8	0.714	1.096
11	2.8	2.9	-0.484	-0.734
12	2.1	2.45	-0.323	-0.491
13	2.2	2.6	-0.260	-0.396
14	2.6	2.6	-0.609	-0.924
15	2.6	2.7	-0.509	-0.773
16	2.9	3.1	-0.371	-0.563
17	5.1	5.1	-0.292	-0.451
18	10.2	10.2	0.355	0.748
19	3.5	3.5	-0.495	-0.751
20	1.2	2.7	0.713	1.100
21	1.7	3	0.577	0.882
22	1.75	2.7	0.233	0.356
23	1.7	2.5	0.077	0.117
24	1.2	2.4	0.413	0.637
25	1.2	4.4	2.413	3.721
26	7.5	7.5	0.013	0.021
27	4.25	4.25	-0.400	-0.610

The standardized residual plot is shown below.

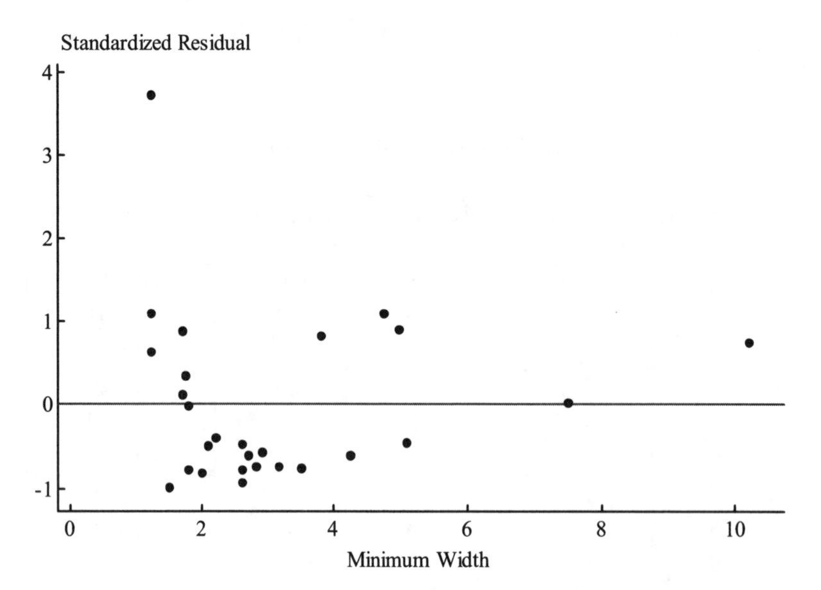

The standardized residual plot shows that there is one point that is a clear outlier (the point whose standardized residual is 3.721). This is the point for product 25.

c The equation of the least-squares regression line is now $\hat{y} = 0.703 + 0.918x$.
A computer analysis gives $s_b = 0.065$. Thus the change in slope from 0.873 to 0.918 expressed in standard deviations is $(0.918 - 0.873)/0.065 = 0.692$. Removal of the point resulted in a reasonably substantial change in the equation of the estimated regression line.

d For every 1-cm increase in minimum width, the mean maximum width is estimated to increase by 0.918 cm.
The intercept would be an estimate of the mean maximum width when the minimum width is zero. It is clearly impossible to have a container whose minimum width is zero.

e

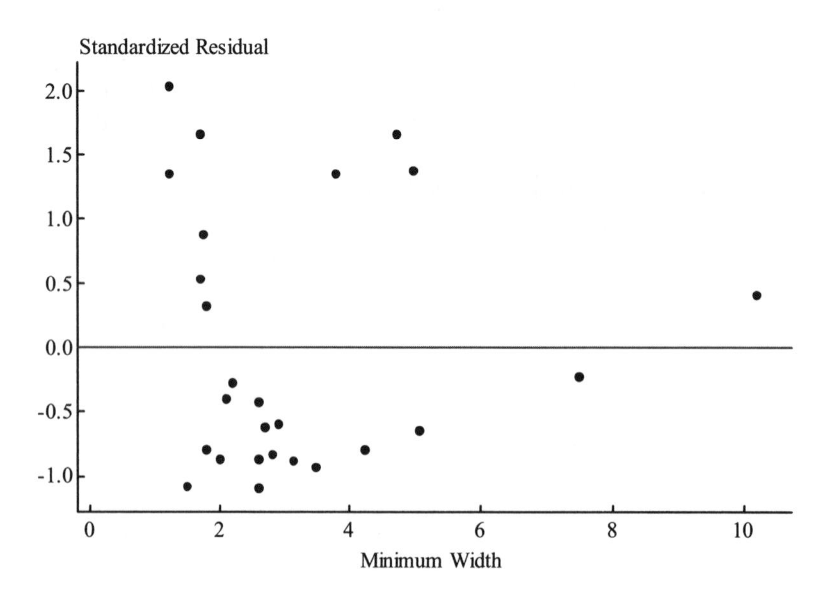

The standardized residual plot for the data with the Coke bottle removed is shown above. The pattern in this plot suggests that the variances of the y distributions decrease as x increases, and therefore that the assumption of constant variance is not valid.

13.31 a

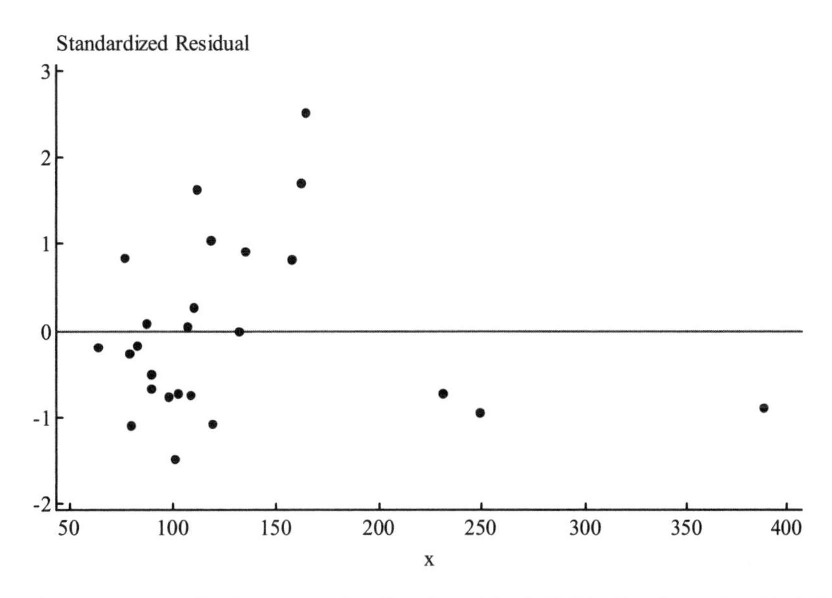

There is one unusually large standardized residual, 2.52, for the point (164.2, 181). The point (387.8, 310) would seem to be an influential point, since removing it from the standardized residual plot would result in an impossible pattern for a residual plot. (The residual plot is likely to be similar in appearance to the standardized residual plot, and the horizontal line at zero should be the least-squares regression line for the residual plot. When the point (387.8, 310) is removed and thus the point with x coordinate 387.8 is removed from the standardized residual plot, the pattern shown amongst the remaining points would seem to result in a least-squares regression line for that plot that has a clearly positive slope.)

b Apart from the one point that has a large residual, the arrangement of points in the residual plot seems consistent with the simple linear regression model.

c If we include the point with the unusually large standardized residual we might begin to suspect that the variances of the y distributions decrease as the x values increase. However, from the relatively small number of points included we do not have particularly strong evidence that the assumption of constant variance does not apply.

13.33 Suppose we constructed a 95% *confidence* interval for the mean value of y when $x = x^*$. We would then be 95% confident that the mean value of y was within that interval. If we were to construct the 95% *prediction* interval at $x = x^*$ we would be 95% confident that an observed y value, y^*, at that value of x will be within the interval. The 95% confidence level for the prediction interval is interpreted as follows. The prediction interval is constructed using a set of independent y values for a given set of x values. Imagine this being done a large number of times, with the prediction interval at $x = x^*$ being calculated for each set of (x, y) points. Imagine, also, a large number of y values being selected at $x = x^*$. Then if one interval is chosen at random, and one y value is chosen at random, on average 95 times out of 100 the y value will be within the interval.

13.35 **a** We use $s_{a+bx^*} = s_e\sqrt{\frac{1}{n}+\frac{(x^*-\bar{x})^2}{S_{xx}}}$.

Here $s_{a+b(2.0)} = 16.486\sqrt{\frac{1}{20}+\frac{(2-2.5)^2}{25}} = \mathbf{4.038}$.

b Since 3 is the same distance from 2.5 as is 2, $s_{a+b(3.0)} = s_{a+b(2.0)} = \mathbf{4.038}$.

c $s_{a+b(2.8)} = 16.486\sqrt{\frac{1}{20}+\frac{(2.8-2.5)^2}{25}} = \mathbf{3.817}$.

d The estimated standard deviation of $a+bx^*$ is smallest for $x^* = 2.5$, since the distance of this value from the mean value of x is zero.

13.37 **a** We need to assume that the conditions for inference are met.
The point estimate of $\alpha+\beta(40)$ is $a+b(40) = 6.843345 - 0.00730608(40) = 6.55110$.
The estimated standard deviation of $a+b(40)$ is

$$s_e\sqrt{\frac{1}{n}+\frac{(40-\bar{x})^2}{S_{xx}}} = 0.0356\sqrt{\frac{1}{16}+\frac{(40-42.375)^2}{7325.75}} = 0.0089547.$$

The critical value of the t distribution with 14 degrees of freedom for a 95% confidence interval is 2.145.
So the required confidence interval is $6.55110 \pm 2.145(0.0089547) = (\mathbf{6.532, 6.570})$.
We are 95% confident that the mean milk pH when the milk temperature is 40°C is between 6.532 and 6.570.

b We need to assume that the conditions for inference are met.
The point estimate of $\alpha+\beta(35)$ is $a+b(35) = 6.843345 - 0.00730608(35) = 6.58763$.
The estimated standard deviation of $a+b(35)$ is

$$s_e\sqrt{\frac{1}{n}+\frac{(35-\bar{x})^2}{S_{xx}}} = 0.0356\sqrt{\frac{1}{16}+\frac{(35-42.375)^2}{7325.75}} = 0.0094138.$$

The critical value of the t distribution with 14 degrees of freedom for a 99% confidence interval is 2.977.
So the required confidence interval is $6.58763 \pm 2.977(0.0094138) = (\mathbf{6.560, 6.616})$.
We are 99% confident that the mean milk pH when the milk temperature is 35°C is between 6.560 and 6.616.

c No. Since 90°C is well outside the range of x values in the original data set, this would not be advisable.

13.39 **a** The equation of the estimated regression line is $\hat{y} = -0.001790 - 0.0021007x$, where x = mean childhood blood lead level and y = brain volume change.

b We need to assume that the conditions for inference are met.
The point estimate of $\alpha+\beta(20)$ is $a+b(20) = -0.001790 - 0.0021007(20) = -0.043804$.
The estimated standard deviation of $a+b(20)$ is

$$s_e\sqrt{\frac{1}{n}+\frac{(20-\bar{x})^2}{S_{xx}}}=0.031\sqrt{\frac{1}{100}+\frac{(20-11.5)^2}{1764}}=0.0069979.$$

The critical value of the t distribution with 98 degrees of freedom for a 90% confidence interval is 1.661.

So the required confidence interval is $-0.043804 \pm 1.661(0.0069979)=(\mathbf{-0.055}, \mathbf{-0.032})$.

We are 90% confident that the mean brain volume change for people with a childhood blood lead level of 20 μg/dL is between −0.055 and −0.032.

c The estimated standard deviation of the amount by which a single y observation deviates from the value predicted by an estimated regression line is

$\sqrt{s_e^2+s_{a+bx^*}^2}=\sqrt{0.031^2+0.0069979^2}=0.03178.$

The critical value of the t distribution with 98 degrees of freedom for a 90% confidence interval is 1.661.

So the required confidence interval is $-0.043804 \pm 1.661(0.03178)=(\mathbf{-0.097}, \mathbf{0.009})$.

We are 90% confident that the brain volume change for a person with a childhood blood lead level of 20 μg/dL will be between −0.097 and 0.009.

d The answer to Part (b) gives an interval in which we are 90% confident that the *mean* brain volume change for a person with a childhood blood lead level of 20 μg/dL lies. The answer to Part (c) states that if we were to find the brain volume change for *one person* with a childhood blood lead level of 20 μg/dL, we are 90% confident that this value will lie within the interval found.

13.41 a The equation of the regression line is $\hat{y}=-133.02+5.919x$, where x = snout vent length and y = clutch size.

b $s_b=\mathbf{1.127}$.

c Yes. Since the estimated slope is positive and since the P-value is small (given as 0.000 in the output) we have convincing evidence that the slope of the population regression line is positive.

d We need to assume that the conditions for inference are met.

The point estimate of $\alpha+\beta(65)$ is $a+b(65)=-133.02+5.919(65)=251.715$.

$S_{xx}=\sum x^2-n\bar{x}^2=45958-14(56.5)^2=1266.5.$

The estimated standard deviation of $a+b(65)$ is

$$s_e\sqrt{\frac{1}{n}+\frac{(65-\bar{x})^2}{S_{xx}}}=33.90\sqrt{\frac{1}{14}+\frac{(65-56.5)^2}{1266.5}}=12.151.$$

Therefore, the estimated standard deviation of the amount by which a single y observation deviates from the value predicted by an estimated regression line is

$\sqrt{s_e^2+s_{a+bx^*}^2}=\sqrt{33.90^2+12.151^2}=36.012.$

The critical value of the t distribution with 12 degrees of freedom for a 95% confidence interval is 2.179.

So the required confidence interval is $251.715 \pm 2.179(36.012)=(\mathbf{173.252}, \mathbf{330.178})$.

We are 95% confident that the clutch size for a salamander whose snout-vent length is 65 will be between 173.252 and 330.178.

e It would not be appropriate to use the estimated regression line to predict the clutch size for a salamander with a snout-vent length of 105, since 105 is a long way outside the range of the x values in the original data set.

13.43 a The equation of the estimated regression line is $\hat{y} = 2.78551 + 0.04462x$, where x = time on shelf and y = moisture content.

b β = slope of the population regression line relating moisture content to shelf time.

H_0: $\beta = 0$

H_a: $\beta \neq 0$

$\alpha = 0.05$

$$t = \frac{b - (\text{hypothesized value})}{s_b} = \frac{b - 0}{s_b}$$

A standardized residual plot is shown below.

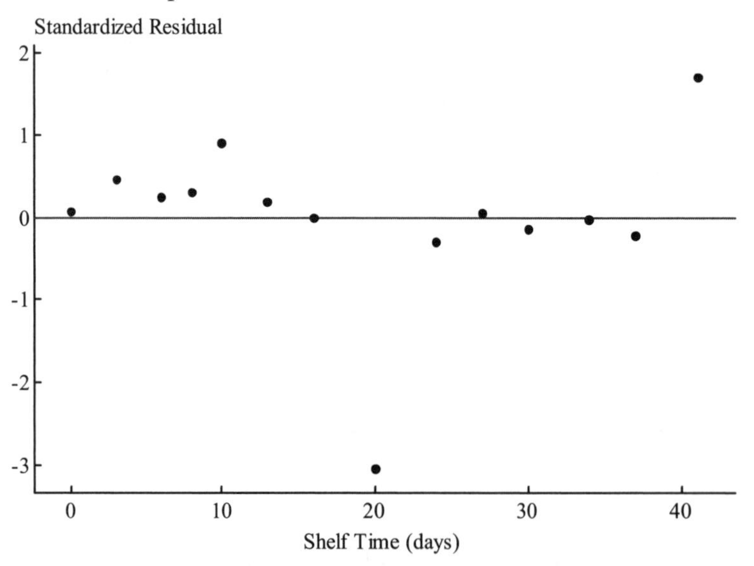

Apart from one outlier, the standardized residual plot shows a random pattern that is consistent with the simple regression model.

$$S_{xx} = \sum x^2 - \frac{\left(\sum x\right)^2}{n} = 7445 - \frac{(269)^2}{14} = 2276.357.$$

$s_e = 0.196246$

$$s_b = \frac{s_e}{\sqrt{S_{xx}}} = 0.00411$$

$$t = \frac{0.04462 - 0}{0.00411} = 10.848$$

df = 12

$P\text{-value} = 2 \cdot P(t_{12} > 10.848) \approx 0$

Since P-value $\approx 0 < 0.05$ we reject H_0. We have convincing evidence that the simple regression model provides useful information for predicting moisture content from knowledge of shelf time.

c The conditions for inference were checked in Part (b).
The point estimate of $\alpha + \beta(30)$ is $a + b(30) = 2.78551 + 0.04462(30) = 4.124$.

$S_{xx} = \sum x^2 - n\bar{x}^2 = 7745 - 14(269/14)^2 = 2576.357.$

$s_e = 0.196246.$

The estimated standard deviation of $a + b(30)$ is

$$s_e\sqrt{\frac{1}{n} + \frac{(30-\bar{x})^2}{S_{xx}}} = 0.196246\sqrt{\frac{1}{14} + \frac{(30-19.214286)^2}{2576.357}} = 0.067006.$$

Therefore, the estimated standard deviation of the amount by which a single y observation deviates from the value predicted by an estimated regression line is

$\sqrt{s_e^2 + s_{a+bx^*}^2} = \sqrt{0.196246^2 + 0.067006^2} = 0.207.$

The critical value of the t distribution with 12 degrees of freedom for a 95% confidence interval is 2.179.
So the required confidence interval is $4.124 \pm 2.179(0.207) = (\mathbf{3.672, 4.576})$.

We are 95% confident that the moisture content for a box of cereal that has been on the shelf for 30 days will be between 3.672 and 4.576 percent.

d Since 4.1 is included in the confidence interval constructed in Part (c), a moisture content exceeding 4.1 percent is quite plausible when the shelf time is 30 days.

13.45 a The scatterplot in Example 5.2 shows a linear pattern that is consistent with the assumptions of the simple linear regression model.
The point estimate of $\alpha + \beta(0.5)$ is $a + b(0.5) = -1.59 + 2.59(0.5) = -0.295$.

$$s_e = \sqrt{\frac{\text{SSResid}}{n-2}} = \sqrt{\frac{1.936}{30}} = 0.254.$$

The estimated standard deviation of $a + b(0.5)$ is

$$s_e\sqrt{\frac{1}{n} + \frac{(0.5-\bar{x})^2}{S_{xx}}} = 0.254\sqrt{\frac{1}{32} + \frac{(0.5-0.6069)^2}{1.479}} = 0.05015.$$

The critical value of the t distribution with 30 degrees of freedom for a 95% confidence interval is 2.042.
So the required confidence interval is $-0.295 \pm 2.042(0.05015) = (\mathbf{-0.397, -0.193})$.

We are 95% confident that the mean perceived astringency score when the tannin concentration is 0.5 is between −0.397 and −0.193.

b We need 95% confidence intervals for the mean astringency ratings at both x values. We already have the confidence interval for $x = 0.5$, so we only need to calculate the interval for $x = 0.7$.
As stated in the solution to Part (a), the scatterplot in Example 5.2 shows a linear pattern that is consistent with the assumptions of the simple linear regression model.
The point estimate of $\alpha + \beta(0.7)$ is $a + b(0.7) = -1.59 + 2.59(0.7) = 0.223$.

$$s_e = \sqrt{\frac{\text{SSResid}}{n-2}} = \sqrt{\frac{1.936}{30}} = 0.254.$$

The estimated standard deviation of $a+b(0.7)$ is

$$s_e\sqrt{\frac{1}{n}+\frac{(0.7-\bar{x})^2}{S_{xx}}}=0.254\sqrt{\frac{1}{32}+\frac{(0.7-0.6069)^2}{1.479}}=0.04894.$$

The critical value of the t distribution with 30 degrees of freedom for a 95% confidence interval is 2.042.

So the required confidence interval is $0.223\pm 2.042(0.04894)=(\mathbf{0.123},\mathbf{0.323})$.

We are 95% confident that the mean perceived astringency score when the tannin concentration is 0.7 is between 0.123 and 0.323.

c The simultaneous confidence level would be $[100-2(1)]\%=98\%$.

d The simultaneous confidence level would be $[100-3(5)]\%=85\%$.

13.47 The statistic r is the correlation coefficient for a sample, while ρ denotes the correlation coefficient for the population.

13.49
1. ρ = the correlation between teaching evaluation index and annual raise for the population from which the sample was selected.
2. H_0: $\rho=0$
3. H_a: $\rho\neq 0$
4. $\alpha=0.05$
5. $t=\dfrac{r}{\sqrt{\dfrac{1-r^2}{n-2}}}$
6. We must assume that the variables have a bivariate normal distribution and that the sample was a random sample from the population.
7. $t=\dfrac{0.11}{\sqrt{\dfrac{1-0.11^2}{351}}}=2.073$
8. df = 351

   $P\text{-value}=2\cdot P(t_{351}>2.073)=0.039$
9. Since $P\text{-value}=0.039<0.05$ we reject H_0. We have convincing evidence of a linear association between teaching evaluation index and annual raise.

This result might be initially surprising, since 0.11 seems to be a relatively small value for the sample correlation coefficient. However, what the result shows is that for a sample size as large as 353, a sample correlation as large as 0.11 would be very unlikely if the population correlation were zero.

13.51 **a**
1. ρ = the correlation between time spent watching television and grade point average for the population from which the sample was selected.
2. H_0: $\rho=0$
3. H_a: $\rho<0$
4. $\alpha=0.01$

5. $t = \dfrac{r}{\sqrt{\dfrac{1-r^2}{n-2}}}$

6. We must assume that the variables have a bivariate normal distribution. We are told that the sample was a random sample.

7. $t = \dfrac{-0.26}{\sqrt{\dfrac{1-(-0.26)^2}{526}}} = -6.175$

8. df = 526
   $P\text{-value} = P(t_{526} < -6.175) \approx 0$

9. Since $P\text{-value} \approx 0 < 0.01$ we reject H_0. We have convincing evidence of a negative correlation between time spent watching television and grade point average.

b Since $r^2 = (-0.26)^2 = 0.0676$, only 6.76% of the observed variation in grade point average would be explained by the regression line. This is not a substantial percentage.

13.53 1. ρ = the correlation between surface and subsurface concentration.

2. H_0: $\rho = 0$

3. H_a: $\rho \neq 0$

4. $\alpha = 0.05$

5. $t = \dfrac{r}{\sqrt{\dfrac{1-r^2}{n-2}}}$

6. We must assume that the sample was a random sample from the population under consideration.

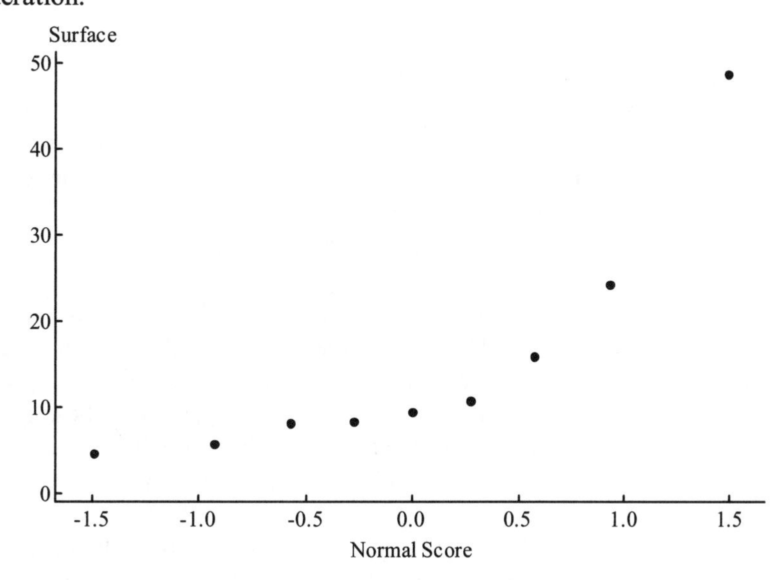

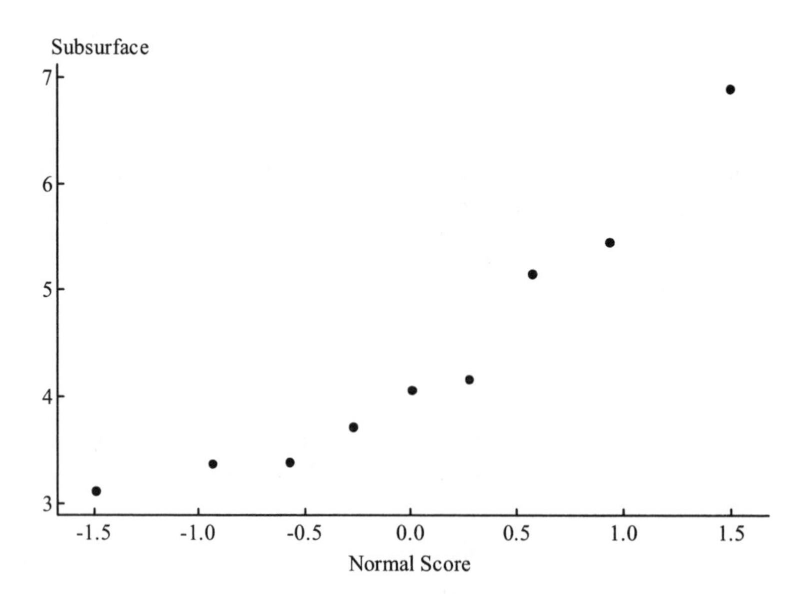

The curved pattern in the first normal probability plot tells us that it is unlikely that the variables have a bivariate normal distribution, but we will nevertheless proceed with the hypothesis test.

7. $r = 0.574$

   $$t = \frac{0.574}{\sqrt{\frac{1-0.574^2}{7}}} = 1.855$$

8. $\text{df} = 7$

   $P\text{-value} = 2 \cdot P(t_7 > 1.855) = 0.106$

9. Since $P\text{-value} = 0.106 > 0.05$ we do not reject H_0. We do not have convincing evidence of a linear relationship between surface and subsurface concentration.

13.55 **a** The slope of the estimated regression line for y = verbal language score against x = height gain from age 11 to 16 is 2.0. This tells us that for each extra inch of height gain the average verbal language score at age 11 increased by 2.0 percentage points. The equivalent results for nonverbal language scores and math scores were 2.3 and 3.0. Thus the reported slopes are consistent with the statement that each extra inch of height gain was associated with an increase in test scores of between 2 and 3 percentage points.

b The slope of the estimated regression line for y = verbal language score against x = height gain from age 16 to 33 is −3.1. This tells us that for each extra inch of height gain the average verbal language score at age 11 decreased by 3.1 percentage points. The equivalent results for nonverbal language scores and math scores were both −3.8. Thus the reported slopes are consistent with the statement that each extra inch of height gain was associated with a decrease in test scores of between 3.1 and 3.8 percentage points.

c Between the ages of 11 and 16 the first boy grew 5 inches more than the second boy. So the first boy's age 11 math score is predicted to be $5 \cdot 3 = 15$ percentage points higher than that of the second boy. Between the ages of 16 and 33 the second boy grew 5 inches more than the first boy. According to this information the first boy's age 11 math score is predicted to be $5 \cdot 3.8 = 19$ percentage points higher than that of the second boy. These two results are consistent with the conclusion that on the whole boys who did their growing early had higher cognitive scores at age 11 than those whose growth occurred later.

13.57 a $t = \dfrac{r}{\sqrt{\dfrac{1-r^2}{n-2}}} = \dfrac{-0.18}{\sqrt{\dfrac{1-(-0.18)^2}{345}}} = -3.399.$

Thus, for a two-tailed test, the *P*-value is $2 \cdot P(t_{345} < -3.399) = 0.001$. Since the *P*-value for a one-tailed test would be a half of this, it is indeed correct, whether this be a one- or two-tailed test, that *P*-value < 0.05.

b Yes. One would expect, generally speaking, that those with greater coping humor ratings would have smaller depression ratings.

c No. Since $r^2 = (-0.18)^2 = 0.0324$, we know that only 3.2% of the variation in depression scale values is attributable to the approximate linear relationship with the coping humor scale. So the linear regression model will generally not give accurate predictions.

13.59 a

1. ρ = the correlation between soil hardness and trail length for the population of penguin burrows.
2. H_0: $\rho = 0$
3. H_a: $\rho < 0$
4. $\alpha = 0.05$
5. $t = \dfrac{r}{\sqrt{\dfrac{1-r^2}{n-2}}}$
6. We must assume that the variables have a bivariate normal distribution and that the sample was a random sample of penguin burrows.
7. $r = -\sqrt{0.386} = -0.621$. (We know that $r < 0$ since the slope of the least-squares line is negative.)
   $$t = \frac{-0.621}{\sqrt{\dfrac{1-(-0.621)^2}{59}}} = -6.090$$
8. df = 59
   $P\text{-value} = P(t_{59} < -6.090) \approx 0$
9. Since $P\text{-value} \approx 0 < 0.05$ we reject H_0. We have convincing evidence of a negative correlation between soil hardness and trail length.

b We need to assume that the conditions for inference are met.
The point estimate of $\alpha + \beta(6.0)$ is $a + b(6.0) = 11.607 - 1.4187(6.0) = 3.0948$.
The estimated standard deviation of $a + b(6.0)$ is

$$s_e\sqrt{\frac{1}{n} + \frac{(6.0-\bar{x})^2}{S_{xx}}} = 2.35\sqrt{\frac{1}{61} + \frac{(6.0-4.5)^2}{250}} = 0.374.$$

Therefore, the estimated standard deviation of the amount by which a single *y* observation deviates from the value predicted by an estimated regression line is

$\sqrt{s_e^2 + s_{a+bx^*}^2} = \sqrt{2.35^2 + 0.374^2} = 2.380.$

The critical value of the t distribution with 59 degrees of freedom for a 95% confidence interval is 2.001.

So the required prediction interval is $3.0948 \pm 2.001(2.380) = (\mathbf{-1.667, 7.856})$.

We are 95% confident that the trail length when the soil hardness is 6.0 will be between −1.667 and 7.856.

c No. For $x = 10$ the least-squares line predicts $y = -2.58$. Since it is not possible to have a negative trail length, it is clear that the simple linear regression model does not apply at $x = 10$. So the simple linear regression model is not suitable for this prediction.

13.61 **a**
1. β = slope of the population regression line relating x = age and y = percentage of the cribriform area of the lamina scleralis occupied by pores.
2. H_0: $\beta = -0.5$
3. H_a: $\beta \neq -0.5$
4. $\alpha = 0.1$
5. $t = \dfrac{b - (\text{hypothesized value})}{s_b} = \dfrac{b - (-0.5)}{s_b}$
6. The data are plotted in the scatterplot below.

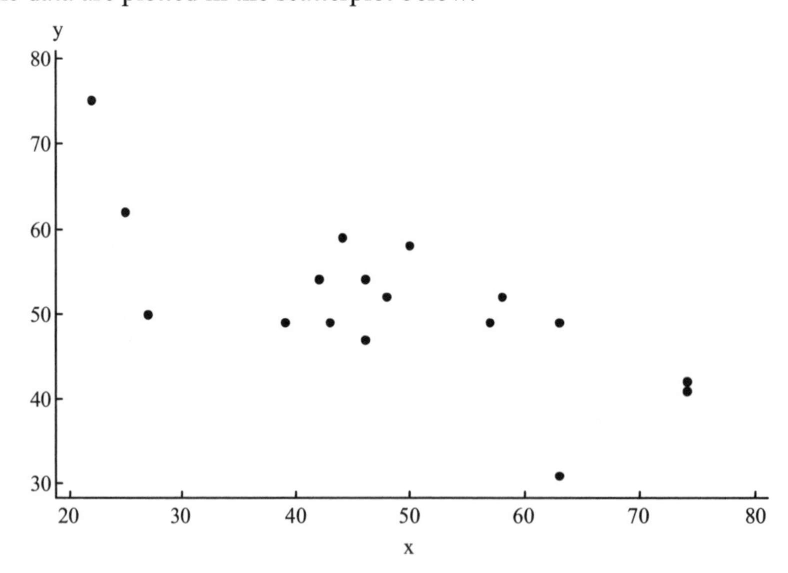

The plot shows a linear pattern, and the vertical spread of points does not appear to be changing over the range of x values in the sample. If we assume that the distribution of errors at any given x value is approximately normal, then the simple linear regression model seems appropriate.

7. $b = -0.447.$

$s_e = 6.75598$

$S_{xx} = 3797.529$

$$s_b = \frac{s_e}{\sqrt{S_{xx}}} = \frac{6.75598}{\sqrt{3797.529}} = 0.1096$$

$$t = \frac{-0.447 - (-0.5)}{0.1096} = 0.488$$

8. df = 15
   $P\text{-value} = 2 \cdot P(t_{15} > 0.488) = 0.633$
9. Since $P\text{-value} = 0.633 > 0.1$ we do not reject H_0. We do not have convincing evidence that the average decrease in percentage area associated with a 1-year age increase is not 0.5.

b As shown in the solution to Part (a), the scatterplot shows a linear pattern that is consistent with the assumptions of the simple linear regression model.
The point estimate of $\alpha + \beta(50)$ is $a + b(50) = 72.918 - 0.447(50) = 50.591$.
The estimated standard deviation of $a + b(50)$ is

$$s_e\sqrt{\frac{1}{n} + \frac{(50-\bar{x})^2}{S_{xx}}} = 6.75598\sqrt{\frac{1}{17} + \frac{(50-48.294)^2}{3797.529}} = 1.649.$$

The critical value of the *t* distribution with 15 degrees of freedom for a 95% confidence interval is 2.131.
So the required confidence interval is $50.591 \pm 2.131(1.649) = \mathbf{(47.076, 54.106)}$.
We are 95% confident that the mean percentage area at age 50 is between 47.076 and 54.106.

13.63 For leptodactylus:
SSResid = 0.30989
Sample size = 9
$b = 0.31636$
$S_{xx} = 42.82$

For bufa:
SSResid = 0.12792
Sample size = 8
$b = 0.35978$
$S_{xx} = 34.54875$

$$s^2 = \frac{0.30989 + 0.12792}{7+6} = 0.03368$$

H_0: $\beta = \beta'$
H_a: $\beta \neq \beta'$
$\alpha = 0.05$

$$t = \frac{b - b'}{\sqrt{\frac{s^2}{S_{xx}} + \frac{s^2}{S'_{xx}}}} = \frac{0.31636 - 0.35978}{\sqrt{\frac{0.03368}{42.82} + \frac{0.03368}{34.54875}}} = -1.03457$$

df = 13
$P\text{-value} = 2 \cdot P(t_{13} < -1.03457) = 0.320$
Since $P\text{-value} = 0.319 > 0.05$ we do not reject H_0. We do not have convincing evidence that the slopes of the population regression lines for the two different frog populations are not equal.

13.65 If the point (20, 33000) is not included, then the slope of the least-squares line would be relatively small and negative (appearing close to horizontal when drawn to the scales of the scatterplot given in the question). If the point is included then the slope of the least-squares line would still be negative, but much further from zero.

13.67 The small *P*-value indicates that there is convincing evidence of a useful linear relationship between percentage raise and productivity.

13.69 **a** The values $e_1,\ldots,e_n$ are the vertical deviations of the y observations from the *population* regression line. The residuals are the vertical deviations from the *sample* regression line.

b False. The simple linear regression model states that the *mean* value of y is equal to $\alpha + \beta x$.

c No. You only test hypotheses about population characteristics; b is a sample statistic.

d Strictly speaking this statement is false, since a set of points lying exactly on a straight line will give a zero result for SSResid. However, it is certainly true to say that, since SSResid is a sum of squares, its value must be *nonnegative*.

e This is not possible, since the sum of the residuals is always zero.

f This is not possible, since SSResid (here said to be equal to 731) is always less than or equal to SSTo (here said to be 615).

Cumulative Review Exercises

CR13.1

Randomly assign the 400 students to two groups of equal size, Group A and Group B. (This can be done by writing the names of the students onto slips of paper, placing the slips into a hat, and picking 200 at random. These 200 people will go into Group A, and the remaining 200 people will go into Group B.) Have the 400 students take the same course, attending the same lectures and being given the same homework assignments. The only difference between the two groups should be that the students in Group A should be given daily quizzes and the students in Group B should not. (This could be done by having the students in Group A take their quizzes in class after the students in Group B have been dismissed.) After the final exam the exam scores for the students in Group A should be compared to the exam scores for the students in Group B.

CR13.3

a Median = 2
Lower quartile = 1.5
Upper quartile = 6.5
IQR = 6.5 − 1.5 = 5

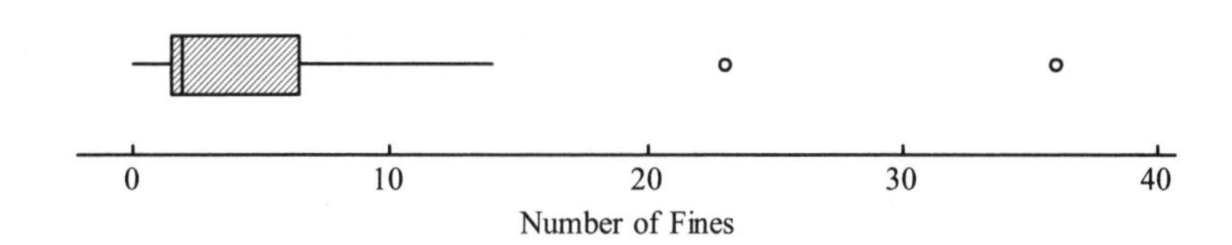

Two of the observations, 23 and 36, are (extreme) outliers.

b The two airlines with the highest numbers of fines assessed may not be the worst in terms of maintenance violations since these airlines might have more flights than the other airlines.

CR13.5

a Check of Conditions

1. Since $n\hat{p} = 1003(0.68) = 682 \geq 10$ and $n(1-\hat{p}) = 1003(0.32) = 321 \geq 10$, the sample size is large enough.
2. The sample size of n = 1003 is much smaller than 10% of the population size (the number of adult Americans).
3. We are told that the survey was nationally representative, so it is reasonable to regard the sample as a random sample from the population of adult Americans.

Calculation

The 95% confidence interval for p is

$$\hat{p} \pm 1.96\sqrt{\frac{\hat{p}(1-\hat{p})}{n}} = 0.68 \pm 1.96\sqrt{\frac{(0.68)(0.32)}{1003}} = (\mathbf{0.651, 0.709}).$$

Interpretation

We are 95% confident that the proportion of all adult Americans who view a landline phone as a necessity is between 0.651 and 0.709.

b 1. p = proportion of all adult Americans who considered a TV set a necessity
2. H_0: $p = 0.5$

3. H_a: $p > 0.5$
4. $\alpha = 0.05$
5. $z = \dfrac{\hat{p} - p}{\sqrt{\dfrac{p(1-p)}{n}}} = \dfrac{\hat{p} - 0.5}{\sqrt{\dfrac{(0.5)(0.5)}{1003}}}$
6. The sample was nationally representative, so it is reasonable to treat the sample as a random sample from the population. The sample size is much smaller than the population size (the number of adult Americans). Furthermore, $np = 1003(0.5) = 501.5 \geq 10$ and $n(1-p) = 1003(0.5) = 501.5 \geq 10$, so the sample is large enough. Therefore the large sample test is appropriate.
7. $z = \dfrac{0.52 - 0.5}{\sqrt{\dfrac{(0.5)(0.5)}{1003}}} = 1.267$
8. $P\text{-value} = P(Z > 1.267) = 0.103$
9. Since $P\text{-value} = 0.103 > 0.05$ we do not reject H_0. We do not have convincing evidence that a majority of adult Americans consider a TV set a necessity.

c 1. p_1 = proportion of adult Americans in 2003 who regarded a microwave oven as a necessity
p_2 = proportion of adult Americans in 2009 who regarded a microwave oven as a necessity
2. H_0: $p_1 - p_2 = 0$
3. H_a: $p_1 - p_2 > 0$
4. $\alpha = 0.01$
5. $z = \dfrac{\hat{p}_1 - \hat{p}_2}{\sqrt{\dfrac{\hat{p}_c(1-\hat{p}_c)}{n_1} + \dfrac{\hat{p}_c(1-\hat{p}_c)}{n_2}}}$
6. We are told that the 2009 survey was nationally representative, so it is reasonable to treat the sample in 2009 as a random sample from the population. We need to assume that the sample in 2003 was a random sample from the population. Also, $n_1\hat{p}_1 = 1003(0.68) = 682 \geq 10$, $n_1(1-\hat{p}_1) = 1003(0.32) = 321 \geq 10$, $n_2\hat{p}_2 = 1003(0.47) = 471 \geq 10$, and $n_2(1-\hat{p}_2) = 1003(0.53) = 532 \geq 10$, so the samples are large enough.
7. $\hat{p}_c = \dfrac{n_1\hat{p}_1 + n_2\hat{p}_2}{n_1 + n_2} = \dfrac{1003(0.68) + 1003(0.47)}{1003 + 1003} = 0.575$

$$z = \frac{0.68 - 0.47}{\sqrt{\dfrac{(0.575)(0.425)}{1003} + \dfrac{(0.575)(0.425)}{1003}}} = 9.513$$
8. $P\text{-value} = P(Z > 9.513) \approx 0$
9. Since $P\text{-value} \approx 0 < 0.01$ we reject H_0. We have convincing evidence that the proportion of adult Americans who regarded a microwave oven as a necessity decreased between 2003 and 2009.

CR13.7

a $P(x = 0) = 1 - 0.38 = \mathbf{0.62}$.

b $P(2 \le x \le 5) = 0.5(0.38) = 0.19.$
$P(x > 5) = 0.18(0.38) = 0.0684.$
So $P(x = 1) = 0.38 - 0.19 - 0.0684 = \mathbf{0.1216}$.

c **0.19**

d **0.0684**

CR13.9

a

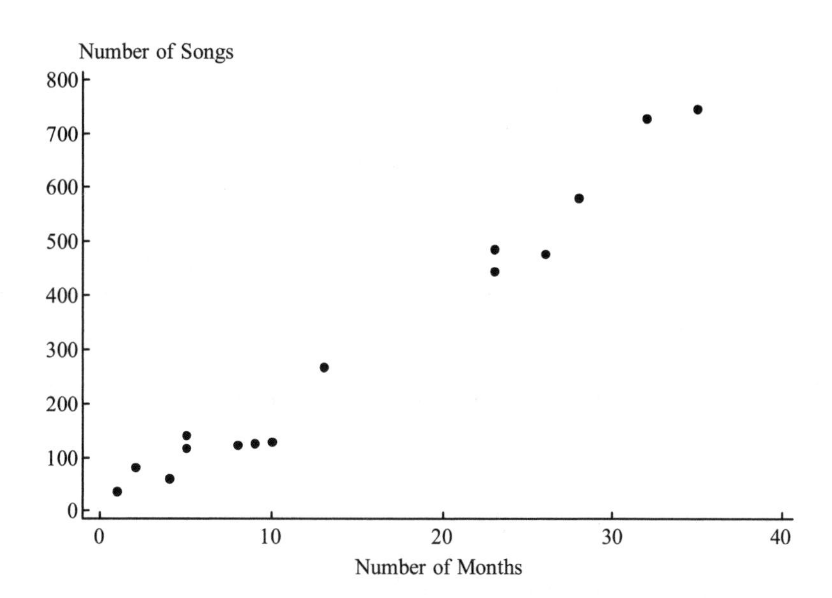

Yes, the relationship looks approximately linear.

b The equation of the estimated regression line is $\hat{y} = -12.887 + 21.126x$, where x = number of months the user has owned the MP3 player and y = number of songs stored.

c

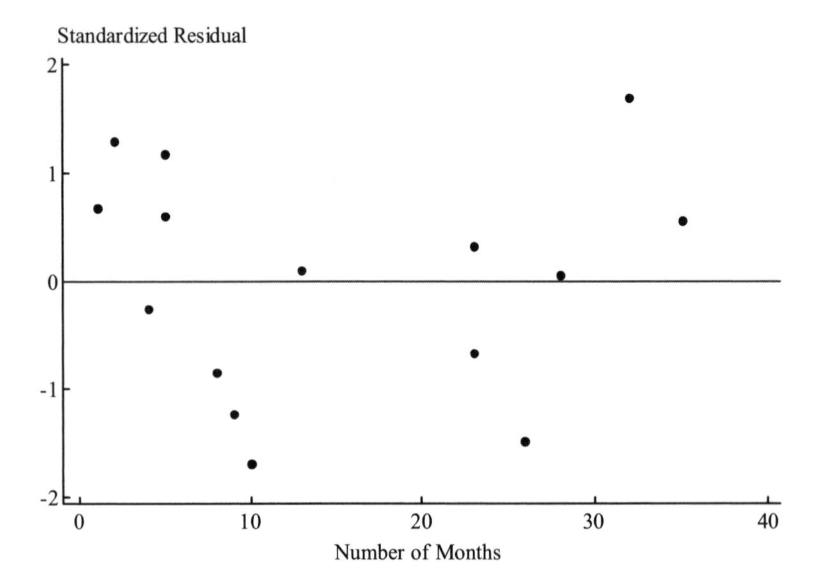

There is a random pattern in the standardized residual plot, and there is no suggestion that the variance of y is not the same at each x value. There are no outliers. The assumptions of the simple linear regression model would therefore seem to be reasonable.

d

1. β = slope of the population regression line relating the number of songs to the number of months.
2. H_0: $\beta = 0$
3. H_a: $\beta \neq 0$
4. $\alpha = 0.05$
5. $t = \dfrac{b - (\text{hypothesized value})}{s_b} = \dfrac{b-0}{s_b}$
6. As explained in Part (c), the assumptions of the simple linear regression model seem to be reasonable.
7. $s_b = 0.994$

   $t = \dfrac{21.126 - 0}{0.994} = 21.263$
8. df = 13

   $P\text{-value} = 2 \cdot P(t_{13} > 21.263) \approx 0$
9. Since $P\text{-value} \approx 0 < 0.05$ we reject H_0. We have convincing evidence of a useful linear relationship between the number of songs stored and the number of months the MP3 player has been owned.

CR13.11

	Year		
Political Affiliation	**2005**	**2004**	**2003**
Democrat	397 (379.706)	409 (375.647)	325 (375.647)
Republican	301 (343.448)	349 (339.776)	373 (339.776)
Independent/Unaffiliated	458 (440.473)	397 (435.764)	457 (435.764)
Other	60 (52.373)	48 (51.813)	48 (51.813)

H_0: The proportions falling in each of the response categories are the same for the three years.
H_a: H_0 is not true.
$\alpha = 0.05$

$$X^2 = \sum_{\text{all cells}} \frac{(\text{observed cell count} - \text{expected cell count})^2}{\text{expected cell count}}$$

The samples were considered to be representative of the populations of undergraduates for the given years, and so it is reasonable to assume that they were random samples from those populations. All the expected counts are greater than 5.

$$X^2 = \frac{(397 - 379.706)^2}{379.706} + \cdots + \frac{(48 - 51.813)^2}{51.813} = 26.175$$

df = 6

$P\text{-value} = P(\chi_6^2 > 26.175) \approx 0$

Since $P\text{-value} \approx 0 < 0.05$ we reject H_0. We have convincing evidence that the distribution of political affiliation is not the same for all three years for which the data are given.

CR13.13

	Credit Card?	
Region	**At Least One Credit Card**	**No Credit Card**
Northeast	401 (429.848)	164 (135.152)
Midwest	162 (150.637)	36 (47.363)
South	408 (397.895)	115 (125.105)
West	104 (96.621)	23 (30.379)

H_0: Region of residence and having a credit card are independent
H_a: Region of residence and having a credit card are not independent
$\alpha = 0.05$

$$X^2 = \sum_{\text{all cells}} \frac{(\text{observed cell count} - \text{expected cell count})^2}{\text{expected cell count}}$$

We are told that the sample was a random sample of undergraduates in the US. All the expected counts are greater than 5.

$$X^2 = \frac{(401 - 429.848)^2}{429.848} + \cdots + \frac{(23 - 30.379)^2}{30.379} = 15.106$$

df = 3

$P\text{-value} = P(\chi_3^2 > 15.106) = 0.002$

Since $P\text{-value} = 0.002 < 0.05$ we reject H_0. We have convincing evidence that region of residence and having a credit card are not independent.

CR13.15

1. μ_1 = mean alkalinity upstream
   μ_2 = mean alkalinity downstream
2. H_0: $\mu_1 - \mu_2 = -50$
3. H_a: $\mu_1 - \mu_2 < -50$
4. $\alpha = 0.05$

5. $t = \dfrac{(\bar{x}_1 - \bar{x}_2) - (\text{hypothesized value})}{\sqrt{\dfrac{s_1^2}{n_1} + \dfrac{s_2^2}{n_2}}} = \dfrac{(\bar{x}_1 - \bar{x}_2) - (-50)}{\sqrt{\dfrac{s_1^2}{n_1} + \dfrac{s_2^2}{n_2}}}$
6. We need to assume that the water specimens were chosen randomly from the two locations. We are given that $n_1 = 24$ and $n_2 = 24$, so neither sample size was greater than or equal to 30. We therefore need to assume that the distributions of alkalinity at the two locations are approximately normal.
7. $t = \dfrac{(75.9 - 183.6) - (-50)}{\sqrt{\dfrac{1.83^2}{24} + \dfrac{1.70^2}{24}}} = -113.169$
8. df $= 45.752$
   $P\text{-value} = P(t_{45.752} < -113.169) \approx 0$
9. Since $P\text{-value} \approx 0 < 0.05$ we reject H_0. We have convincing evidence that the mean alkalinity is higher downstream than upstream by more than 50 mg/L.

CR13.17

Direction	Observed Count	Expected Count
0° to <45°	12	15
45° to < 90°	16	15
90° to <135°	17	15
135° to <180°	15	15
180° to <225°	13	15
225° to <270°	20	15
270° to <315°	17	15
315° to <360°	10	15

1. Let $p_1, \ldots, p_8$ be the proportions of homing pigeons choosing the twelve given directions.
2. H_0: $p_1 = \cdots = p_8 = 0.125$
3. H_a: H_0 is not true
4. $\alpha = 0.1$
5. $X^2 = \sum_{\text{all cells}} \dfrac{(\text{observed cell count} - \text{expected cell count})^2}{\text{expected cell count}}$
6. We need to assume that the study was performed using a random sample of homing pigeons. All the expected counts are greater than 5.
7. $X^2 = \dfrac{(12-15)^2}{15} + \cdots + \dfrac{(10-15)^2}{15} = 4.8$
8. df $= 7$
   $P\text{-value} = P(\chi_7^2 > 4.8) = 0.684$
9. Since $P\text{-value} = 0.684 > 0.1$ we do not reject H_0. We do not have convincing evidence that the birds exhibit a preference.

Chapter 14
Multiple Regression Analysis

Note: In this chapter, numerical answers to questions involving the normal, *t*, chi square, and *F* distributions were found using values from a calculator. Students using statistical tables will find that their answers differ slightly from those given.

14.1 An example of a deterministic model is $y = \alpha + \beta_1 x_1 + \beta_2 x_2 + \beta_3 x_3$. This is a deterministic model, because, for any given values of x_1, x_2, and x_3, the value of *y* is known. An example of a probabilistic model is $y = \alpha + \beta_1 x_1 + \beta_2 x_2 + \beta_3 x_3 + e$. The error term, *e*, is a random variable: we do not know what value it is going to take. Consequently, *y*, too, is a random variable.

14.3 **a** The population regression function is $30 + 0.90x_1 + 0.08x_2 - 4.50x_3$.

b The population regression coefficients are 0.90, 0.08, and −4.50.

c When dynamic hand grip endurance and trunk extension ratio are fixed, the mean increase in rating of acceptable load associated with a 1-cm increase in extent of left lateral bending is 0.90 kg.

d When extent of left lateral bending and dynamic hand grip endurance are fixed, the mean decrease in rating of acceptable load associated with a 1-N/kg increase in trunk extension ratio is 4.50 kg.

e Mean of $y = 30 + 0.90(25) + 0.08(200) - 4.50(10) = \mathbf{23.5}$ kg.

f For these values of the independent variables, the distribution of *y* is normal, with mean 23.5 and standard deviation 5. We require

$$P(13.5 < y < 33.5) = P\left(\frac{13.5 - 23.5}{5} < z < \frac{33.5 - 23.5}{5}\right) = P(-2 < z < 2) = \mathbf{0.9545}.$$

14.5 **a** When $x_1 = 20$ and $x_2 = 50$, mean weight $= -21.658 + 0.828(20) + 0.373(50) = 13.552$ g.

b When length is fixed, the mean increase in weight associated with a 1-mm increase in width is 0.828 g.
When width is fixed, the mean increase in weight associated with a 1-mm increase in length is 0.373 g.

14.7 **a** Mean yield $= 415.11 - 6.6(20) - 4.5(40) = \mathbf{103.11}$.

b Mean yield $= 415.11 - 6.6(18.9) - 4.5(43) = \mathbf{96.87}$.

c When the average percentage of sunshine is fixed, the mean decrease in yield associated with a 1-degree increase in average temperature is 6.60.
When the average temperature is fixed, the mean decrease in yield associated with a one percentage point increase in average percentage of sunshine is 4.50.

14.9 **a**

x	2	4	6	8	10	12
Mean of y	354	456	526	564	570	544

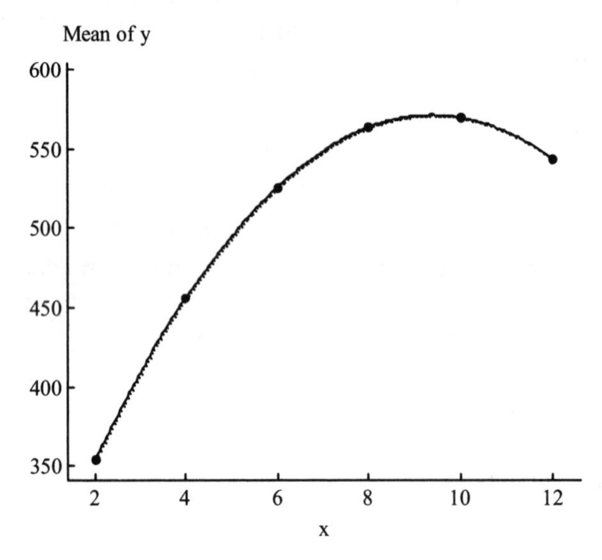

b The values calculated in Part (a) show us that the chlorine content is greater for a degree of delignification value of 10 than for a degree of delignification value of 8.

c When $x = 9$, mean of $y = 571$.
When degree of delignification increases from 8 to 9, mean chlorine content increases by 7.
When degree of delignification increases from 9 to 10, mean chlorine content decreases by 1.

14.11 **a**

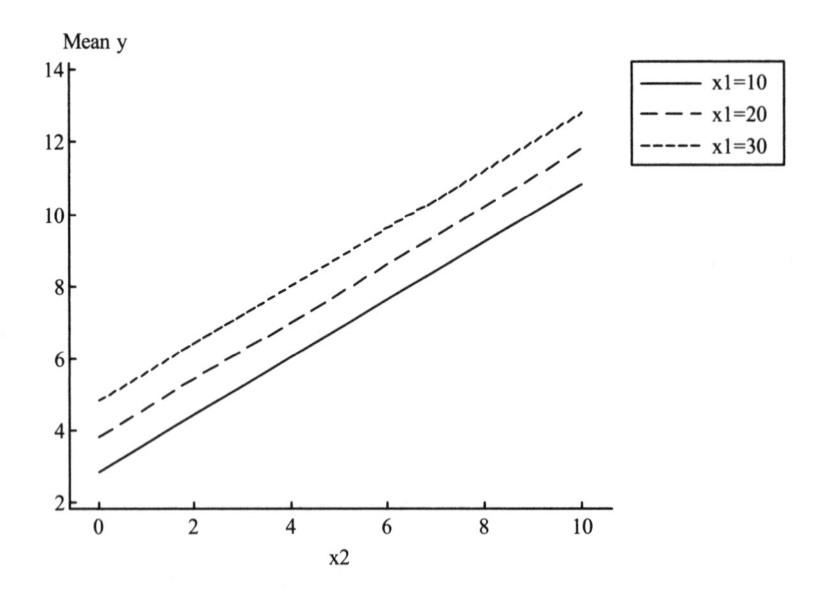

b

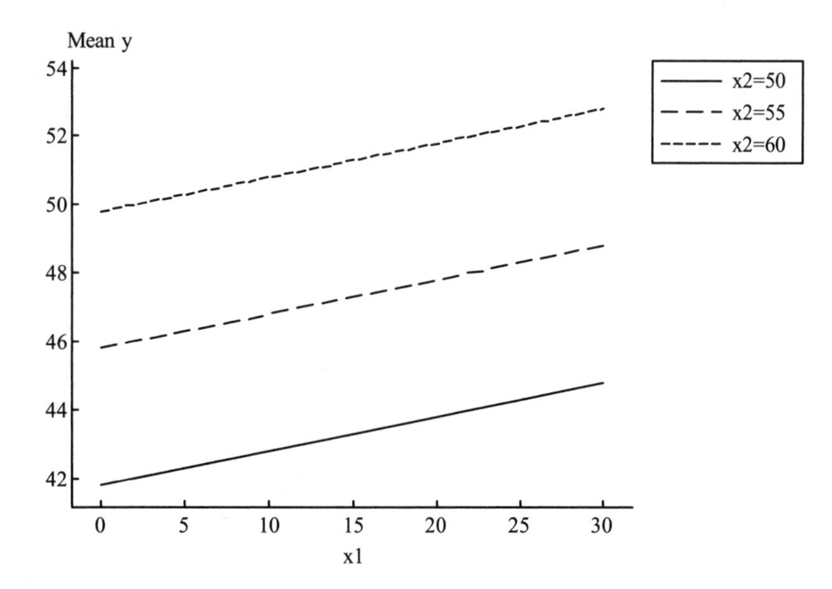

c The fact that there is no interaction between x_1 and x_2 is reflected by the fact that in each of the graph, the lines are parallel.

d

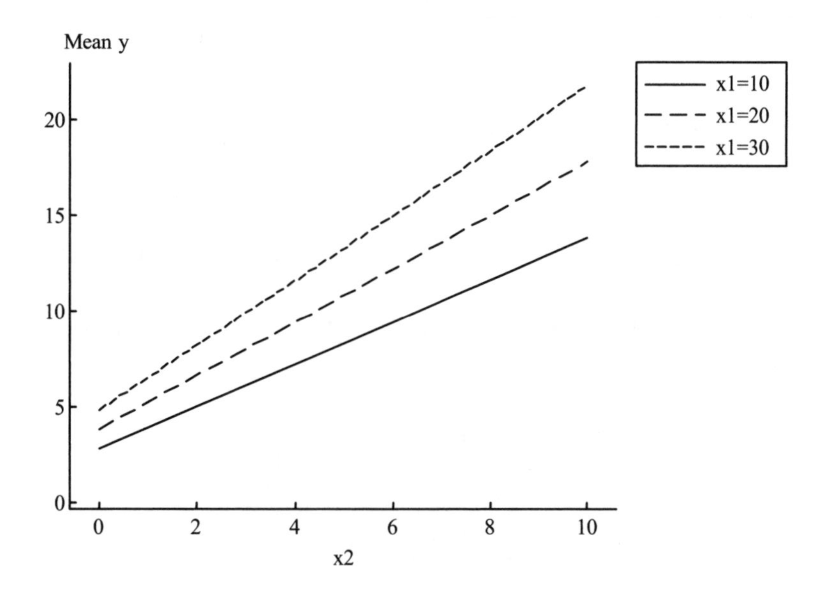

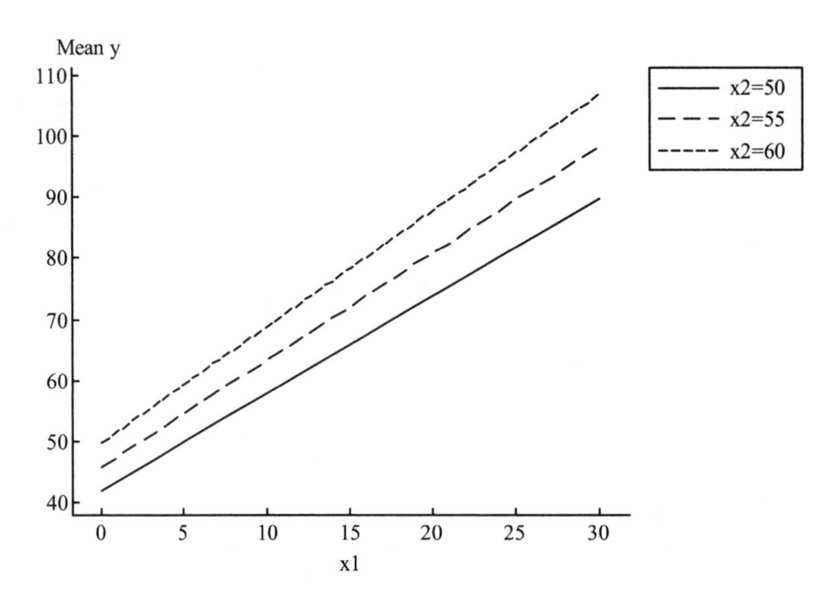

The presence of an interaction term causes the lines in the graphs to be nonparallel.

14.13 **a** $y = \alpha + \beta_1 x_1 + \beta_2 x_2 + \beta_3 x_3 + e$

b $y = \alpha + \beta_1 x_1 + \beta_2 x_2 + \beta_3 x_3 + \beta_4 x_1^2 + \beta_5 x_2^2 + \beta_6 x_3^2 + e$

c $y = \alpha + \beta_1 x_1 + \beta_2 x_2 + \beta_3 x_3 + \beta_4 x_2 x_3 + e$
$y = \alpha + \beta_1 x_1 + \beta_2 x_2 + \beta_3 x_3 + \beta_4 x_1 x_3 + e$
$y = \alpha + \beta_1 x_1 + \beta_2 x_2 + \beta_3 x_3 + \beta_4 x_1 x_2 + e$

d $y = \alpha + \beta_1 x_1 + \beta_2 x_2 + \beta_3 x_3 + \beta_4 x_1^2 + \beta_5 x_2^2 + \beta_6 x_3^2 + \beta_7 x_2 x_3 + \beta_8 x_1 x_3 + \beta_9 x_1 x_2 + e$

14.15 **a** We need additional variables x_3, x_4, and x_5. The values of these variables could be defined as shown in the table.

Size Class	x_3	x_4	x_5
Subcompact	0	0	0
Compact	1	0	0
Midsize	0	1	0
Large	0	0	1

The model equation is $y = \alpha + \beta_1 x_1 + \beta_2 x_2 + \beta_3 x_3 + \beta_4 x_4 + \beta_5 x_5 + e$

b The additional predictors are $x_1 x_3$, $x_1 x_4$, and $x_1 x_5$.

14.17 **a** $P\text{-value} = P(F_{3,15} > 4.23) = \mathbf{0.024}$.

b $P\text{-value} = P(F_{4,18} > 1.95) = \mathbf{0.146}$.

c $P\text{-value} = P(F_{5,20} > 4.10) = \mathbf{0.010}$.

d P-value $= P(F_{4,35} > 4.58) = \mathbf{0.004}$.

14.19 **a**
1. The model is $y = \alpha + \beta_1 x_1 + \beta_2 x_2 + \beta_3 x_3 + e$ where y = surface area, x_1 = weight, x_2 = width, and x_3 = length.
2. H_0: $\beta_1 = \beta_2 = \beta_3 = 0$
3. H_a: At least one of the β_i's is not zero.
4. $\alpha = 0.05$
5. $F = \dfrac{R^2/k}{(1-R^2)/(n-(k+1))}$
6. Since we do not have the original data set we are unable to check the conditions. We need to assume that the variables are related according to the model given above, and that the random deviations, e, are normally distributed with mean zero and fixed standard deviation.
7. $F = \dfrac{0.996/3}{(1-0.996)/146} = 12118$
8. P-value $= P(F_{3,146} > 12118) \approx 0$
9. Since P-value $\approx 0 < 0.05$ we reject H_0. We have convincing evidence that the multiple regression model is useful.

b Since the P-value is small and r^2 is close to 1 there is strong evidence that the model is useful.

c The model in Part (b) should be recommended, since adding the variables x_1 and x_2 to the model (to obtain the model in Part (a)) only increases the value of R^2 a small amount (from 0.994 to 0.996).

14.21
1. The model is $y = \alpha + \beta_1 x_1 + \beta_2 x_2 + \beta_3 x_3 + \beta_4 x_4 + \beta_5 x_5 + \beta_6 x_6 + e$, where y = species richness, x_1 = watershed area, x_2 = shore width, x_3 = drainage, x_4 = water color, x_5 = sand percentage, and x_6 = alkalinity.
2. H_0: $\beta_1 = \beta_2 = \beta_3 = \beta_4 = \beta_5 = \beta_6 = 0$
3. H_a: At least one of the β_i's is not zero.
4. $\alpha = 0.01$
5. $F = \dfrac{R^2/k}{(1-R^2)/(n-(k+1))}$
6. Since we do not have the original data set we are unable to check the conditions. We need to assume that the variables are related according to the model given above, and that the random deviations, e, are normally distributed with mean zero and fixed standard deviation.
7. $F = \dfrac{0.83/6}{(1-0.83)/30} = 24.412$
8. P-value $= P(F_{6,30} > 24.412) \approx 0$
9. Since P-value $\approx 0 < 0.01$ we reject H_0. We have convincing evidence that the chosen model is useful.

14.23 1. The model is $y = \alpha + \beta_1 x_1 + \beta_2 x_2 + \beta_3 x_3 + \beta_4 x_4 + e$, where y = fish intake, x_1 = water temperature, x_2 = number of pumps running, x_3 = sea state, and x_4 = speed.

2. H_0: $\beta_1 = \beta_2 = \beta_3 = \beta_4 = 0$
3. H_a: At least one of the β_i 's is not zero.
4. $\alpha = 0.1$
5. $F = \dfrac{\text{SSRegr}/k}{\text{SSResid}/(n-(k+1))}$
6. Since we do not have the original data set we are unable to check the conditions. We need to assume that the variables are related according to the model given above, and that the random deviations, e, are normally distributed with mean zero and fixed standard deviation.
7. $F = \dfrac{1486.9/4}{2230.2/21} = 3.500$
8. $P\text{-value} = P(F_{4,21} > 3.500) = 0.024$
9. Since $P\text{-value} = 0.024 < 0.1$ we reject H_0. We have convincing evidence that the model is useful.

14.25 1. The model is $y = \alpha + \beta_1 x_1 + \beta_2 x_2 + \beta_3 x_3 + \beta_4 x_4 + \beta_5 x_5 + \beta_6 x_6 + \beta_7 x_7 + \beta_8 x_8 + \beta_9 x_9 + e$, where y = ecology score, x_1 = age times 10, x_2 = income, x_3 = gender, x_4 = race, x_5 = number of years of education, x_6 = ideology, x_7 = social class, x_8 = postmaterialist (0 or 1), and x_9 = materialist (0 or 1).

2. H_0: $\beta_1 = \beta_2 = \beta_3 = \beta_4 = \beta_5 = \beta_6 = \beta_7 = \beta_8 = \beta_9 = 0$
3. H_a: At least one of the β_i 's is not zero.
4. $\alpha = 0.05$
5. $F = \dfrac{R^2/k}{(1-R^2)/(n-(k+1))}$
6. Since we do not have the original data set we are unable to check the conditions. We need to assume that the variables are related according to the model given above, and that the random deviations, e, are normally distributed with mean zero and fixed standard deviation.
7. We have $n = 1136$ and $k = 9$. So $F = \dfrac{0.06/9}{(1-0.06)/1126} = 7.986$.
8. $P\text{-value} = P(F_{9,1126} > 7.986) \approx 0$
9. Since $P\text{-value} \approx 0 < 0.05$ we reject H_0. We have convincing evidence that the multiple regression model is useful.

14.27 **a** The MINITAB output is shown below.

```
The regression equation is
Catch Time = 1.44 - 0.0523 Prey Length + 0.00397 Prey Speed


Predictor          Coef     SE Coef       T      P
Constant        1.43958     0.08325   17.29  0.000
Prey Length    -0.05227     0.01459   -3.58  0.002
Prey Speed    0.0039700   0.0006194    6.41  0.000


S = 0.0930752   R-Sq = 75.0%   R-Sq(adj) = 71.9%
```

```
Analysis of Variance

Source          DF       SS        MS       F      P
Regression       2  0.41617  0.20809  24.02  0.000
Residual Error  16  0.13861  0.00866
Total           18  0.55478
```

The estimated regression equation is $\hat{y} = 1.43958 - 0.05227x_1 + 0.0039700x_2$, where y = catch time, x_1 = prey length, and x_2 = prey speed.

b When $x_1 = 6$ and $x_2 = 50$, $\hat{y} = 1.43958 - 0.05227(6) + 0.0039700(50) = \mathbf{1.324}$ seconds.

c
1. The model is $y = \alpha + \beta_1 x_1 + \beta_2 x_2 + e$, with the variables as defined above.
2. H_0: $\beta_1 = \beta_2 = 0$
3. H_a: At least one of the β_i's is not zero.
4. $\alpha = 0.05$
5. $F = \dfrac{R^2/k}{(1-R^2)/(n-(k+1))}$
6. The normal probability plot of the standardized residuals is shown below.

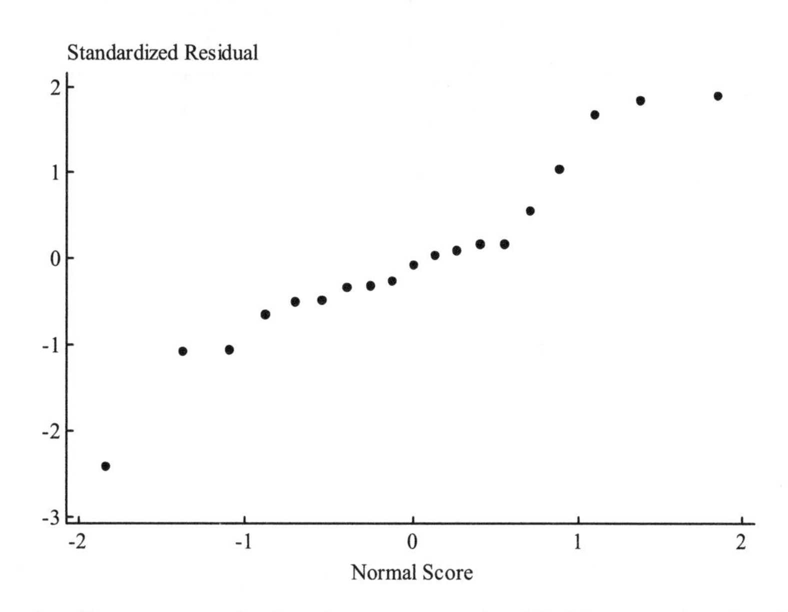

   There is a linear pattern in the plot, so we are justified in assuming that the random deviations are normally distributed.
7. $F = 24.02$
8. P-value $= 0.000$.
9. Since P-value $= 0.000 < 0.05$ we reject H_0. We have convincing evidence that the multiple regression model is useful for predicting catch time.

d The values of the new variable are shown in the table below.

Prey Length	Prey Speed	Catch Time	x
7	20	1.10	0.3500
6	20	1.20	0.3000
5	20	1.23	0.2500
4	20	1.40	0.2000
3	20	1.50	0.1500
3	40	1.40	0.0750
4	40	1.36	0.1000
6	40	1.30	0.1500
7	40	1.28	0.1750
7	80	1.40	0.0875
6	60	1.38	0.1000
5	80	1.40	0.0625
7	100	1.43	0.0700
6	100	1.43	0.0600
7	120	1.70	0.0583
5	80	1.50	0.0625
3	80	1.40	0.0375
6	100	1.50	0.0600
3	120	1.90	0.0250

The MINITAB output is shown below.

```
The regression equation is
Catch Time = 1.59 - 1.40 x


Predictor     Coef   SE Coef      T      P
Constant   1.58648   0.04803  33.03  0.000
x          -1.4044    0.3124  -4.50  0.000


S = 0.122096   R-Sq = 54.3%   R-Sq(adj) = 51.6%


Analysis of Variance

Source          DF       SS       MS      F      P
Regression       1  0.30135  0.30135  20.22  0.000
Residual Error  17  0.25342  0.01491
Total           18  0.55478
```

The estimated regression equation is $\hat{y} = 1.58648 - 1.4044x$.

e Since both the R^2 and the adjusted R^2 values shown in the computer outputs are greater for the first model than for the second, the first model is preferable to the second. The first model is the one that accounts for the greater proportion of the observed variation in catch time.

14.29 a SSResid = **390.435**, SSTo = **1618.209**, SSRegr = 1618.209 − 390.435 = **1227.775**.

b $R^2 = 1 - \frac{\text{SSResid}}{\text{SSTo}} = 1 - \frac{390.435}{1618.209} = \mathbf{0.759}$.

This tells us that 75.9% of the observed variation in shear strength can be explained by the fitted model.

c

1. The model is $y = \alpha + \beta_1 x_1 + \beta_2 x_2 + \beta_3 x_3 + \beta_4 x_4 + \beta_5 x_5 + e$, where y = shear strength, x_1 = depth, x_2 = water content, $x_3 = x_1^2$, $x_4 = x_2^2$, and $x_5 = x_1 x_2$.
2. H_0: $\beta_1 = \beta_2 = \beta_3 = \beta_4 = \beta_5 = 0$
3. H_a: At least one of the β_i 's is not zero.
4. $\alpha = 0.05$
5. $F = \frac{R^2/k}{(1-R^2)/(n-(k+1))}$
6. The normal probability plot of the standardized residuals is shown below.

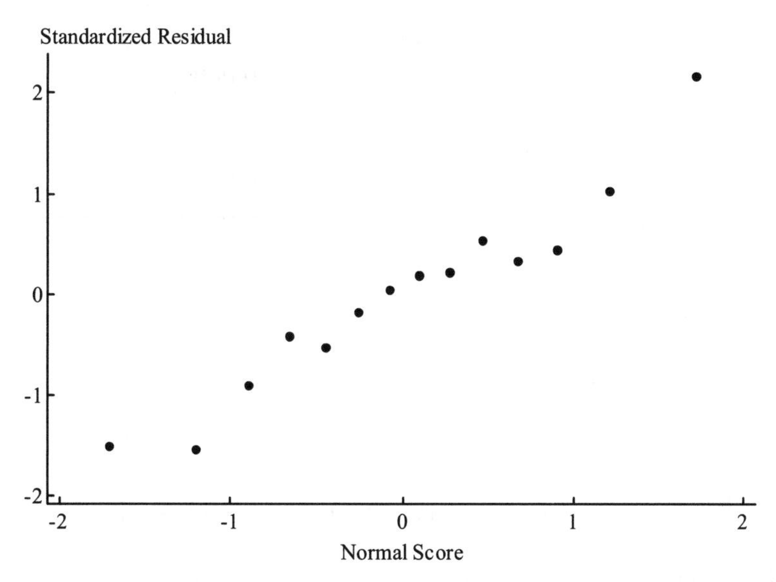

The plot shows a linear pattern, so we are justified in assuming that the random deviations are normally distributed.

7. $F = \frac{0.759/5}{0.241/8} = 5.031$
8. $P\text{-value} = P(F_{5,8} > 5.031) = 0.022$
9. Since $P\text{-value} = 0.022 < 0.05$ we reject H_0. We have convincing evidence that the multiple regression model is useful.

14.31

1. The model is $y = \alpha + \beta_1 x_1 + \beta_2 x_2 + e$, where y = yield, x_1 = defoliation level, and $x_2 = x_1^2$.
2. H_0: $\beta_1 = \beta_2 = 0$
3. H_a: At least one of the β_i 's is not zero.
4. $\alpha = 0.01$
5. $F = \frac{R^2/k}{(1-R^2)/(n-(k+1))}$

6. Since we do not have the original data set we are unable to check the conditions. We need to assume that the variables are related according to the model given above, and that the random deviations, *e*, are normally distributed with mean zero and fixed standard deviation.
7. $F = \dfrac{0.902/2}{(1-0.902)/21} = 96.643$
8. $P\text{-value} = P(F_{2,21} > 96.643) \approx 0$
9. Since $P\text{-value} \approx 0 < 0.01$ we reject H_0. We have convincing evidence that the quadratic model specifies a useful relationship between y and x.

14.33 The MINITAB output is shown below.

```
The regression equation is
y = - 151 - 16.2 x1 + 13.5 x2 + 0.0935 x3 - 0.253 x4 + 0.492 x5


Predictor       Coef  SE Coef      T      P
Constant      -151.4    134.1  -1.13  0.292
x1           -16.216    8.831  -1.84  0.104
x2            13.476    8.187   1.65  0.138
x3           0.09353  0.07093   1.32  0.224
x4           -0.2528   0.1271  -1.99  0.082
x5            0.4922   0.2281   2.16  0.063


S = 6.98783   R-Sq = 75.9%   R-Sq(adj) = 60.8%


Analysis of Variance

Source          DF       SS      MS     F      P
Regression       5  1227.57  245.51  5.03  0.022
Residual Error   8   390.64   48.83
Total           13  1618.21
```

This verifies the estimated regression equation given in the question.

14.35 The MINITAB output is shown below.

```
The regression equation is
y = 35.8 - 0.68 x1 + 1.28 x2


Predictor     Coef  SE Coef      T      P
Constant     35.83    53.54   0.67  0.508
x1          -0.676    1.436  -0.47  0.641
x2          1.2811   0.4243   3.02  0.005


S = 22.9789   R-Sq = 55.0%   R-Sq(adj) = 52.1%


Analysis of Variance

Source          DF     SS     MS      F      P
Regression       2  20008  10004  18.95  0.000
Residual Error  31  16369    528
Total           33  36377
```

The estimated regression equation is $\hat{y} = 35.83 - 0.676x_1 + 1.2811x_2$, where y = infestation rate, x_1 = mean temperature, and x_2 = mean relative humidity.

1. The model is $y = \alpha + \beta_1 x_1 + \beta_2 x_2 + e$, with the variables defined as above.
2. H_0: $\beta_1 = \beta_2 = 0$
3. H_a: At least one of the β_i 's is not zero.
4. $\alpha = 0.05$
5. $F = \dfrac{R^2/k}{(1-R^2)/(n-(k+1))}$
6. The normal probability plot of the standardized residuals is shown below.

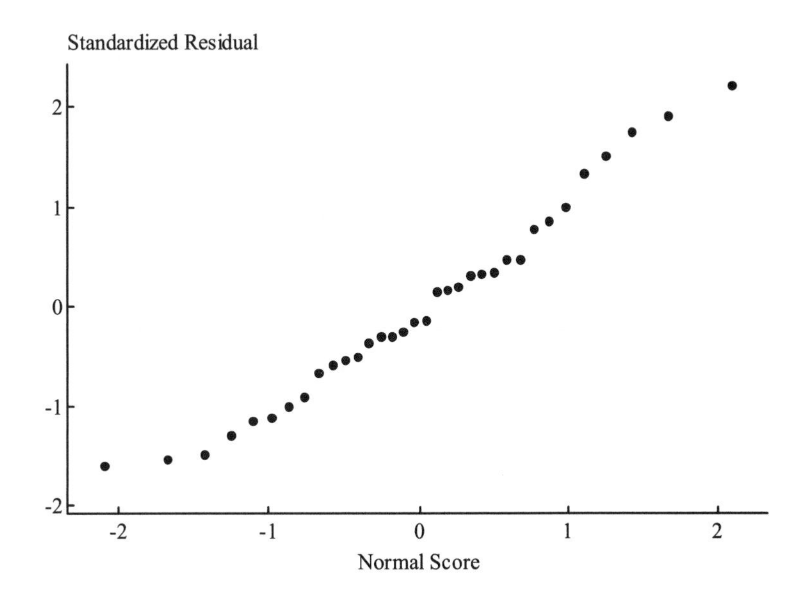

The plot shows a linear pattern, so we are justified in assuming that the random deviations are normally distributed.

7. $F = 18.95$
8. P-value $= 0.000$
9. Since $P\text{-value} = 0.000 < 0.05$ we reject H_0. We have convincing evidence that the multiple regression model is useful.

Chapter 15
Analysis of Variance

Note: In this chapter, numerical answers to questions involving the normal, t, chi square, and F distributions were found using values from a calculator. Students using statistical tables will find that their answers differ slightly from those given.

15.1 **a** $P\text{-value} = P(F_{4,15} > 5.37) = \mathbf{0.007}$.

b $P\text{-value} = P(F_{4,15} > 1.90) = \mathbf{0.163}$.

c $P\text{-value} = P(F_{4,15} > 4.89) = \mathbf{0.010}$.

d $P\text{-value} = P(F_{3,20} > 14.48) = \mathbf{0.000}$.

e $P\text{-value} = P(F_{3,20} > 2.69) = \mathbf{0.074}$.

f $P\text{-value} = P(F_{4,50} > 3.24) = \mathbf{0.019}$.

15.3 **a** Let $\mu_1, \mu_2, \mu_3, \mu_4$ be the mean lengths of stay for people participating in the four health plans.
H_0: $\mu_1 = \mu_2 = \mu_3 = \mu_4$
H_a: At least two among $\mu_1, \mu_2, \mu_3, \mu_4$ are different.

b $\text{df}_1 = k - 1 = 3,\ \text{df}_2 = N - k = 32 - 4 = 28$.
$P\text{-value} = P(F_{3,28} > 4.37) = \mathbf{0.012}$.
Since $P\text{-value} = 0.012 > 0.01$ we do not reject H_0. We do not have convincing evidence that mean length of stay is related to health plan.

c $\text{df}_1 = k - 1 = 3,\ \text{df}_2 = N - k = 32 - 4 = 28$.
$P\text{-value} = P(F_{3,28} > 4.37) = \mathbf{0.012}$.
Since $P\text{-value} = 0.012 > 0.01$ we do not reject H_0. We do not have convincing evidence that mean length of stay is related to health plan.

15.5 Summary statistics are given in the table below.

	7+ label	12+ label	16+ label	18+ label
n	10	10	10	10
$\bar{x}$	4.8	6.8	7.1	8.1
s	2.098	1.619	1.524	1.449
s^2	4.4	2.622	2.322	2.1

1. Let $\mu_1, \mu_2, \mu_3, \mu_4$ be the mean ratings for the four restrictive rating labels.
2. H_0: $\mu_1 = \mu_2 = \mu_3 = \mu_4$

3. H_a: At least two among μ_1, μ_2, μ_3, μ_4 are different.
4. $\alpha = 0.05$
5. $F = \dfrac{\text{MSTr}}{\text{MSE}}$
6. Boxplots for the four groups are shown below.

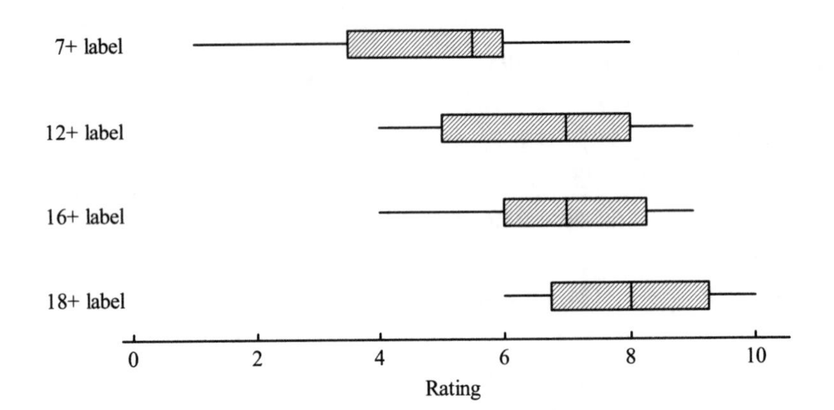

The boxplots are close enough to being symmetrical, and there are no outliers. The largest standard deviation (2.098) is not more than twice the smallest (1.449). We are told to assume that the boys were randomly assigned to the four age label ratings.

7. $N = 10 + 10 + 10 + 10 = 40$
Grand total $= 10(4.8) + 10(6.8) + 10(7.1) + 10(8.1) = 268$

$$\bar{\bar{x}} = \frac{268}{40} = 6.7$$

$$\begin{aligned}\text{SSTr} &= n_1(\bar{x}_1 - \bar{\bar{x}})^2 + n_2(\bar{x}_2 - \bar{\bar{x}})^2 + n_3(\bar{x}_3 - \bar{\bar{x}})^2 + n_4(\bar{x}_4 - \bar{\bar{x}})^2 \\ &= 10(4.8 - 6.7)^2 + 10(6.8 - 6.7)^2 + 10(7.1 - 6.7)^2 + 10(8.1 - 6.7)^2 \\ &= 57.4\end{aligned}$$

Treatment df $= k - 1 = 3$

$$\begin{aligned}\text{SSE} &= (n_1 - 1)s_1^2 + (n_2 - 1)s_2^2 + (n_3 - 1)s_3^2 + (n_4 - 1)s_4^2 \\ &= 9(4.4) + 9(2.622) + 9(2.322) + 9(2.1) \\ &= 103\end{aligned}$$

Error df $= N - k = 40 - 4 = 36$

$$F = \frac{\text{MSTr}}{\text{MSE}} = \frac{\text{SSTr/treatment df}}{\text{SSE/error df}} = \frac{57.4/3}{103/36} = 6.687$$

8. $P\text{-value} = P(F_{3,36} > 6.687) = 0.001$
9. Since $P\text{-value} = 0.001 < 0.05$ we reject H_0. We have convincing evidence that the mean ratings for the four restrictive rating labels are not all equal.

15.7 Summary statistics are shown in the table below.

	Treatment 1	Treatment 2	Treatment 3	Treatment 4
n	18	25	17	14
$\bar{x}$	5.778	6.480	3.529	2.929
s	4.081	3.441	2.401	2.129
s^2	16.654	11.843	5.765	4.533

1. Let $\mu_1, \mu_2, \mu_3, \mu_4$ be the mean numbers of pretzels consumed for the four treatments.
2. H_0: $\mu_1 = \mu_2 = \mu_3 = \mu_4$
3. H_a: At least two among $\mu_1, \mu_2, \mu_3, \mu_4$ are different.
4. $\alpha = 0.05$
5. $F = \dfrac{\text{MSTr}}{\text{MSE}}$
6. Boxplots for the four groups are shown below.

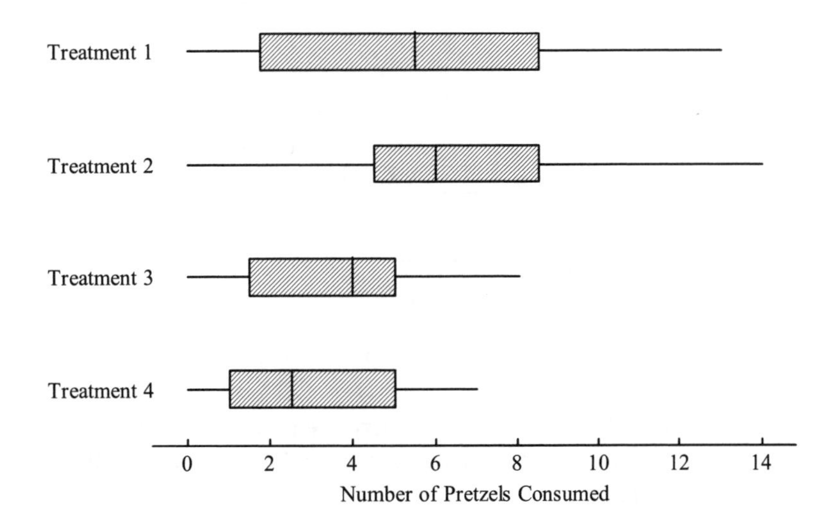

The boxplots are roughly symmetric, and there are no outliers. The largest standard deviation (4.081) is not more than twice the smallest (2.129). We are told that the men were randomly assigned to the four treatments.

7. $N = n_1 + n_2 + n_3 + n_4 = 74$

   Grand total $= n_1\bar{x}_1 + n_2\bar{x}_2 + n_3\bar{x}_3 + n_4\bar{x}_4 = 367$

   $\bar{\bar{x}} = \dfrac{\text{grand total}}{N} = 4.959$

   $\text{SSTr} = n_1(\bar{x}_1 - \bar{\bar{x}})^2 + n_2(\bar{x}_2 - \bar{\bar{x}})^2 + n_3(\bar{x}_3 - \bar{\bar{x}})^2 + n_4(\bar{x}_4 - \bar{\bar{x}})^2$
   $= 162.363$

   Treatment df $= k - 1 = 3$

   $\text{SSE} = (n_1 - 1)s_1^2 + (n_2 - 1)s_2^2 + (n_3 - 1)s_3^2 + (n_4 - 1)s_4^2$
   $= 718.515$

Error df $= N - k = 70$

$$F = \frac{\text{MSTr}}{\text{MSE}} = \frac{\text{SSTr/treatment df}}{\text{SSE/error df}} = 5.273$$

8. P-value $= P(F_{3,70} > 5.273) = 0.002$
9. Since P-value $= 0.002 < 0.05$ we reject H_0. We have convincing evidence that the mean numbers of pretzels consumed for the four treatments are not all equal.

15.9

1. Let μ_1, μ_2, μ_3, μ_4 be the mean changes in body fat mass for the four treatments.
2. H_0: $\mu_1 = \mu_2 = \mu_3 = \mu_4$
3. H_a: At least two among μ_1, μ_2, μ_3, μ_4 are different.
4. $\alpha = 0.05$
5. $F = \dfrac{\text{MSTr}}{\text{MSE}}$
6. Boxplots for the four groups are shown below.

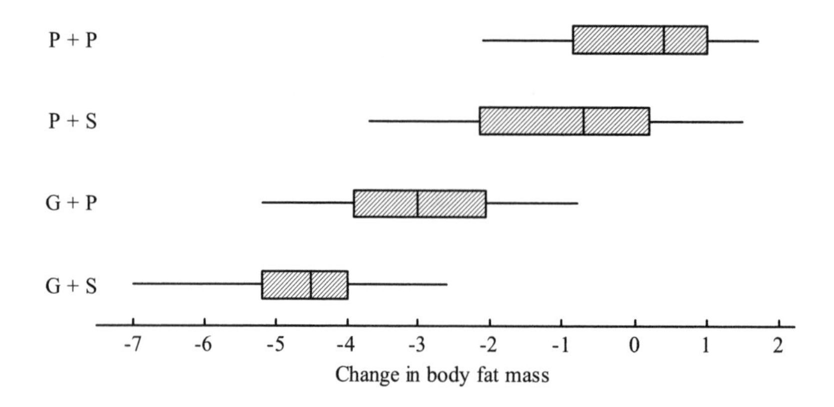

The boxplots are roughly symmetric, and there are no outliers. The largest standard deviation (1.443) is not more than twice the smallest (1.122). We are told that the men were randomly assigned to the four treatments.

7. $N = 74$

Grand total $= -158.3$

$$\bar{\bar{x}} = \frac{\text{grand total}}{N} = -2.139$$

$$\text{SSTr} = n_1(\bar{x}_1 - \bar{\bar{x}})^2 + n_2(\bar{x}_2 - \bar{\bar{x}})^2 + n_3(\bar{x}_3 - \bar{\bar{x}})^2 + n_4(\bar{x}_4 - \bar{\bar{x}})^2$$
$$= 247.403$$

Treatment df $= k - 1 = 3$

$$\text{SSE} = (n_1 - 1)s_1^2 + (n_2 - 1)s_2^2 + (n_3 - 1)s_3^2 + (n_4 - 1)s_4^2$$
$$= 107.314$$

Error df $= N - k = 70$

$$F = \frac{\text{MSTr}}{\text{MSE}} = \frac{\text{SSTr/treatment df}}{\text{SSE/error df}} = 53.793$$

8. P-value $= P(F_{3,70} > 53.793) \approx 0$
9. Since P-value $\approx 0 < 0.05$ we reject H_0. We have convincing evidence that the mean change in body fat differs for the four treatments.

15.11
1. Let μ_1, μ_2, μ_3 be the mean Hopkins scores for the three populations.
2. H_0: $\mu_1 = \mu_2 = \mu_3$
3. H_a: At least two among μ_1, μ_2, μ_3 are different.
4. $\alpha = 0.05$
5. $F = \dfrac{\text{MSTr}}{\text{MSE}}$
6. We are told to treat the samples as random samples from their respective populations. We have to assume that the population Hopkins score distributions are approximately normal with the same standard deviation.
7. $N = n_1 + n_2 + n_3 = 234$

   Grand total $= n_1\bar{x}_1 + n_2\bar{x}_2 + n_3\bar{x}_3 = 7064.66$

   $\bar{\bar{x}} = \dfrac{\text{grand total}}{N} = 30.191$

   $\text{SSTr} = n_1(\bar{x}_1 - \bar{\bar{x}})^2 + n_2(\bar{x}_2 - \bar{\bar{x}})^2 + n_3(\bar{x}_3 - \bar{\bar{x}})^2$
   $= 100.786$

   Treatment df $= k - 1 = 2$

   $\text{SSE} = (n_1 - 1)s_1^2 + (n_2 - 1)s_2^2 + (n_3 - 1)s_3^2 + (n_4 - 1)s_4^2$
   $= 4409.036$

   Error df $= N - k = 231$

   $F = \dfrac{\text{MSTr}}{\text{MSE}} = \dfrac{\text{SSTr/treatment df}}{\text{SSE/error df}} = 2.640$
8. $P\text{-value} = P(F_{2,231} > 2.640) = 0.074$
9. Since $P\text{-value} = 0.074 > 0.05$ we do not reject H_0. We do not have convincing evidence that the mean Hopkins scores are not the same for all three student populations.

15.13 $k = 4$, $N = 20$. Treatment df $= k - 1 = 3$. Error df $= N - k = 16$.
$\text{SSTr} = \text{SSTo} - \text{SSE} = 310500.76 - 235419.04 = 75081.72$.
The completed table is shown below.

Source of Variation	df	Sum of Squares	Mean Square	F
Treatments	3	75081.72	25027.24	1.701
Error	16	235419.04	14713.69	
Total	19	310500.76		

1. Let μ_1, μ_2, μ_3, μ_4 be the mean number of miles until failure for the four given brands of spark plug.
2. H_0: $\mu_1 = \mu_2 = \mu_3 = \mu_4$
3. H_a: At least two among μ_1, μ_2, μ_3, μ_4 are different.
4. $\alpha = 0.05$
5. $F = \dfrac{\text{MSTr}}{\text{MSE}}$
6. We need to treat the samples as random samples from their respective populations, and assume that the population distributions are approximately normal with the same standard deviation.

7. $F = 1.701$
8. $P\text{-value} = P(F_{3,16} > 1.701) = 0.207$
9. Since $P\text{-value} = 0.207 > 0.05$ we do not reject H_0. We do not have convincing evidence that the mean number of miles to failure is not the same for all four brands of spark plug.

15.15 Since there is a significant difference in all three of the pairs we need a set of intervals none of which includes zero. **Set 3** is therefore the required set.

15.17 a In *decreasing* order of the resulting mean numbers of pretzels eaten the treatments were: slides with related text, slides with no text, slides with unrelated text, and no slides. There were no significant differences between the results for slides with no text and slides with unrated text, and for slides with unrelated text and no slides. However there was a significant difference between the results for slides with related text and each one of the other treatments, and between the results for no slides and for slides with no text (and for slides with related text).

b The results for the women and men are almost exactly the reverse of one another, with, for example, slides with related text (treatment 2) resulting in the smallest mean number of pretzels eaten for the women and the largest mean number of pretzels eaten for the men. For the men, treatment 2 was significantly different from all the other treatments; however for women treatment 2 was not significantly different from treatment 1. For both women and men there was a significant difference between treatments 1 and 4 and no significant difference between treatments 3 and 4. However, between treatments 1 and 3 there was a significant difference for the women but no significant difference for the men.

15.19 a

	Driving	**Shooting**	**Fighting**
Sample mean	3.42	4.00	5.30

(Driving and Shooting are underlined together.)

b

	Driving	**Shooting**	**Fighting**
Sample mean	2.81	3.44	4.01

(Driving and Shooting are underlined together; Shooting and Fighting are underlined together.)

15.21 a $N = n_1 + n_2 + n_3 + n_4 = 80$

Grand total $= n_1\bar{x}_1 + n_2\bar{x}_2 + n_3\bar{x}_3 + n_4\bar{x}_4 = 158$

$$\bar{\bar{x}} = \frac{\text{grand total}}{N} = 1.975$$

$$\text{SSTr} = n_1(\bar{x}_1 - \bar{\bar{x}})^2 + n_2(\bar{x}_2 - \bar{\bar{x}})^2 + n_3(\bar{x}_3 - \bar{\bar{x}})^2 + n_4(\bar{x}_4 - \bar{\bar{x}})^2$$
$$= 13.450$$

Treatment df $= k - 1 = 3$

$$\text{SSE} = (n_1 - 1)s_1^2 + (n_2 - 1)s_2^2 + (n_3 - 1)s_3^2 + (n_4 - 1)s_4^2$$
$$= 7.465$$

Error df $= N - k = 76$

$$F = \frac{\text{MSTr}}{\text{MSE}} = \frac{\text{SSTr/treatment df}}{\text{SSE/error df}} = 45.644$$

The ANOVA table is shown below.

Source of Variation	df	Sum of Squares	Mean Square	F
Treatments	3	13.450	4.483	45.644
Error	76	7.465	0.098	
Total	79	20.915		

1. Let $\mu_1, \mu_2, \mu_3, \mu_4$ be the mean numbers of seeds germinating for the four treatments.
2. H_0: $\mu_1 = \mu_2 = \mu_3 = \mu_4$
3. H_a: At least two among $\mu_1, \mu_2, \mu_3, \mu_4$ are different.
4. $\alpha = 0.05$
5. $F = \dfrac{\text{MSTr}}{\text{MSE}}$
6. We need to assume that the samples of 100 seeds collected from each treatment were random samples from those populations, and that the population distributions of numbers of seeds germinating are approximately normal with the same standard deviation.
7. $F = 45.644$
8. $P\text{-value} = P(F_{3,76} > 45.644) \approx 0$
9. Since $P\text{-value} \approx 0 < 0.05$ we reject H_0. We have convincing evidence that the mean number of seeds germinating is not the same for all four treatments.

b We will construct the T-K interval for $\mu_2 - \mu_3$.
Appendix Table 7 gives the 95% Studentized range critical value $q = 3.74$ (using $k = 4$ and error df $= 60$, the closest tabled value to $\text{df} = n - k = 76$). The T-K interval for $\mu_2 - \mu_3$ is

$$(2.35 - 1.70) \pm 3.74\sqrt{\frac{0.098}{2}\left(\frac{1}{20} + \frac{1}{20}\right)} = (0.388, 0.912).$$

Since this interval does not contain zero, we have convincing evidence that seeds eaten and then excreted by lizards germinate at a different rate from those eaten and then excreted by birds. Therefore, since the sample mean was higher for the lizard dung treatment than for the bird dung treatment, we have convincing evidence that seeds eaten and then excreted by lizards germinate at a *higher* rate from those eaten and then excreted by birds.